AF595046

Faunistical and Ecological Studies on Carabid Beetles

Faunistical and Ecological Studies on Carabid Beetles

Editor

Tibor Magura

MDPI • Basel • Beijing • Wuhan • Barcelona • Belgrade • Manchester • Tokyo • Cluj • Tianjin

Editor
Tibor Magura
University of Debrecen
Hungary

Editorial Office
MDPI
St. Alban-Anlage 66
4052 Basel, Switzerland

This is a reprint of articles from the Special Issue published online in the open access journal *Diversity* (ISSN 1424-2818) (available at: https://www.mdpi.com/journal/diversity/special_issues/faunistical_ecological_carabid_beetles).

For citation purposes, cite each article independently as indicated on the article page online and as indicated below:

LastName, A.A.; LastName, B.B.; LastName, C.C. Article Title. *Journal Name* **Year**, *Volume Number*, Page Range.

ISBN 978-3-0365-3102-1 (Hbk)
ISBN 978-3-0365-3103-8 (PDF)

Cover image courtesy of Tibor Magura

© 2022 by the authors. Articles in this book are Open Access and distributed under the Creative Commons Attribution (CC BY) license, which allows users to download, copy and build upon published articles, as long as the author and publisher are properly credited, which ensures maximum dissemination and a wider impact of our publications.
The book as a whole is distributed by MDPI under the terms and conditions of the Creative Commons license CC BY-NC-ND.

Contents

About the Editor

Tibor Magura (Ph.D., D.Sc.) is Professor at the Department of Ecology, Faculty of Science and Technology, University of Debrecen. His main research interests include urban ecology, forest ecology, agroecology and conservation biology, with special emphasis on ground beetles.

Preface to "Faunistical and Ecological Studies on Carabid Beetles"

Carabid beetles (Coleoptera: Carabidae) are one of the best-known taxa in entomology. They are common in most terrestrial habitats and geographical areas and can easily be collected using standard methods. Taxonomy, phylogeny, geographic distribution, habitat associations, and ecological requirements of carabid species are well known. Based on the above, carabids are often used as model organisms in faunistical and ecological studies. Original research papers published in the Special Issue "Faunistical and Ecological Studies on Carabid Beetles" of the journal Diversity covered the aspects of carabid faunistic and ecology.

Tibor Magura
Editor

Article

Habitat and Landform Types Drive the Distribution of Carabid Beetles at High Altitudes

Mauro Gobbi [1,*], Marco Armanini [2], Teresa Boscolo [1], Roberta Chirichella [3], Valeria Lencioni [1], Simone Ornaghi [1] and Andrea Mustoni [2]

[1] Section of Invertebrate Zoology and Hydrobiology, MUSE-Science Museum, Corso del Lavoro e della Scienza 3, 38122 Trento, Italy; tb.entomologo@gmail.com (T.B.); Valeria.Lencioni@muse.it (V.L.); simone.ornaghi@studenti.unimi.it (S.O.)

[2] Adamello-Brenta Natural Park, Via Nazionale 12, 38080 Strembo, Italy; marcoarmanini@pnab.it (M.A.); andreamustoni@pnab.it (A.M.)

[3] Department of Veterinary Medicine, University of Sassari, Via Vienna 2, 07100 Sassari, Italy; roberta.chirichella@gmail.com

* Correspondence: mauro.gobbi@muse.it

Abstract: The high altitude mountain slopes of the Dolomites (Italian Alps) are characterized by great habitat and landform heterogeneities. In this paper, we investigated the effect of Nature 2000 habitat and landform types in driving the high altitude ground beetle (Carabidae) distribution in the Western Dolomites (Brenta group, Italy). We studied the carabid assemblages collected in 55 sampling points distributed in four Nature 2000 habitat types and four landform types located between 1860 and 2890 m above sea level (a.s.l.). Twenty-two species, half of them Alpine endemics, were sampled. Species richness and taxonomic distinctness did not show any significant difference among habitat types; conversely, these differences became significant when the landform type was considered. Total activity density and the frequency of brachypterous, endemic and predatory species showed significant differences between both habitat and landform types. Indicator species analysis identified twelve species linked to a specific habitat type and thirteen species linked to a specific landform type. Canonical correspondence analysis showed that altitude and vegetation cover drove the species distribution in each habitat and landform type while the aspect had a weak effect. Our results highlight the need for a geomorphological characterization of the sampling points when high altitude ground-dwelling arthropods are investigated.

Keywords: Alps; Dolomites; ground beetles; geomorphology; Nature 2000; rocky landforms; species distribution

Citation: Gobbi, M.; Armanini, M.; Boscolo, T.; Chirichella, R.; Lencioni, V.; Ornaghi, S.; Mustoni, A. Habitat and Landform Types Drive the Distribution of Carabid Beetles at High Altitudes. *Diversity* **2021**, *13*, 142. https://doi.org/10.3390/d13040142

Academic Editors: Luc Legal and Tibor Magura

Received: 25 February 2021
Accepted: 24 March 2021
Published: 26 March 2021

Publisher's Note: MDPI stays neutral with regard to jurisdictional claims in published maps and institutional affiliations.

Copyright: © 2021 by the authors. Licensee MDPI, Basel, Switzerland. This article is an open access article distributed under the terms and conditions of the Creative Commons Attribution (CC BY) license (https://creativecommons.org/licenses/by/4.0/).

1. Introduction

High altitude areas experience high landscape complexity driven not only by the occurrence of different habitat types but also by the different landforms on the mountain slopes that can increase the number of microhabitats available for colonization by plants and/or arthropods [1,2].

Carabid beetles (Coleoptera: Carabidae), along with springtails (Collembola), spiders (Arachnida: Araneae) and myriapods (Chilopoda and Diplopoda) are among the most common and most studied ground-dwelling arthropods living at high altitudes [2–5]. High mountain ecosystems especially above the treeline are mainly governed by climatic factors; therefore, changes in the occurrence of alpine species and in the composition of their assemblages are highly relevant for monitoring the impacts of climate change on the biotic component [2–6].

An increasing number of recent studies have investigated spatial and temporal patterns in species richness and species life history traits along altitudinal gradients [6–11]. However, information about the composition and distribution of species assemblages in the mosaic of habitat and rocky landform types available at a high altitude is still lacking for most of the mountain regions. This is probably due to the difficulties in reaching a few high altitude areas, in working under harsh environmental conditions and identifying in detail the different kind of landforms, specifically in alpine fellfields. However, our expectation is that recording both habitat and landform type will increase our understanding of carabid distribution in the mountains because landform diversity within each habitat type gives more detailed information about the availability of microhabitats determined by the geomorphological features of the slope.

Quantitative studies on the assemblage characteristics and habitat specificity of carabid beetles at high elevations (i.e., above the treeline) in the European Alps are available only for a few specific habitats such as glacial and periglacial landforms [2] and specific mountain groups such as Paneveggio-Pale di San Martino and Sarapiss (Eastern Dolomites, Italy) [12,13].

As the mountain group of the Dolomites (Italian Alps) is one of the most spectacular examples of the result of geomorphological processes molding the mountain slopes [14], we selected it as the study area. Specifically, it offers a high geomorphological heterogeneity within each Nature 2000 habitat type. Nature 2000 is the largest coordinated network of protected areas in the world, aimed at protecting Europe's species and habitat diversity; more than 230 rare and characteristic habitat types, which must be protected, have been identified.

The aim of this study was to investigate the carabid beetle distribution in Nature 2000 habitats and rocky landforms located above the treeline of the Western Dolomites (Italian Alps). Specifically, we tested whether (1) carabid beetles' richness, taxonomic distinctness, activity density and percentage of high altitude specialists were associated with specific Nature 2000 habitats and landforms and (2) whether species occurrence was linked to specific habitat or landform types. Due to the great habitat and landform heterogeneity of the mountain slopes of the Dolomites [14] our expectation was to find a synergic effect of the Nature 2000 habitat and landform types in driving high altitude carabid beetle distribution.

Our study is the first to apply a geomorphological approach (i.e., the study of landform types) in addition to the habitat type approach to document carabid beetle distributions at high altitudes.

2. Materials and Methods

The Dolomites are a mountain range in the northern Italian Alps, which rise to above 3000 m above sea level (a.s.l.) and cover 141,903 ha. They are a UNESCO World Heritage Site and they are of international significance for geomorphology because they present a wide range of landforms related to erosion, tectonism and glaciation (https://whc.unesco.org/en/list/1237/, accessed on 1 March 2021).

A total of 11 plots were selected in the mountain group of the Brenta Dolomites (the westernmost mountain group of the Dolomites, Adamello-Brenta Natural Park) (Figure 1). Each plot was designed to represent the habitat and the landform heterogeneity available starting from the uppermost larch (*Larix decida*) forest (about 1900 m a.s.l.) to the Cima Grosté mountain peak (2901 m a.s.l.). Specifically, six plots were located along the north-west slope of the Brenta group (Vallesinella Valley, investigated in 2018) and five along the north-east slope (Tovel Valley, investigated in 2019).

In each plot (ca. 314 m^2) five pitfall traps were placed 50 m apart to avoid spatial dependency. Each pitfall trap consisted of a plastic vessel (diameter 7 cm, height 10 cm) baited with a mixture of wine-vinegar and salt [2]. The traps were active over the entire snow-free season from early July to late September and collected and reset ca. every 20 days.

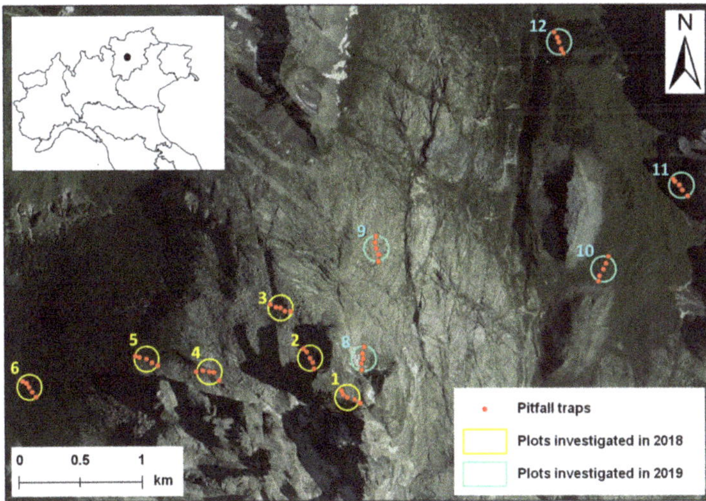

Figure 1. Study area located in the Adamello-Brenta Natural Park, Western Dolomites, Italian Alps (46°11′1.94″ N, 10°54′2.41″ E; in black in the left panel) with the 11 investigated plots and the 55 sampling points (pitfall traps).

The following information was available for each trap: altitude (m a.s.l.), vegetation cover (%), Nature 2000 habitat code and landform type. These data were derived from the Digital Elevation Model (DEM) raster at a 10 m spatial resolution (PAT Geoportal: http://www.territorio.provincia.tn.it, accessed on 10 February 2021), the Habitat Nature 2000 code vector map (PAT Geoportal: http://www.territorio.provincia.tn.it, accessed on 10 February 2021) and the landform type vector map (Geomorphologic database of BioMiti Project [15] in QGIS 3.10 [16].

A total of 55 sampling points (pitfall traps), ranging from ca. 1860 to ca. 2900 m a.s.l., were analyzed separately. For each trap the following information was obtained: species richness (i.e., the total number of species collected during the sampling season), abundance (i.e., the total number of individuals sampled), taxonomic distinctness (i.e., a measure that emphasizes the average taxonomic relatedness between species within a community [17,18]) and the percentage of species within the community sampled by the trap with the following species traits occurring simultaneously: brachypterous + endemic + predator (hereafter BEP species). BEP species are those with life history traits typical of the species living at high altitude; specifically, a low dispersal ability, restricted distributional range and specialized diet [10,12,19].

Carabids were identified to the species level following Pesarini and Monzini [20,21] and are preserved in the entomological collection at the MUSE-Science Museum of Trento (Italy).

Traps at a higher altitude were kept active for a shorter period than those located at a lower altitude due to the shorter snow-free period. Therefore, to obtain comparable data among traps, the number of carabid individuals collected in each trap was transformed to an activity density (AD) [22] using the following formula: AD = (no. of individuals/(days of trap activity)). For each trap, the total activity density (TotalAD) was calculated by summing up the AD values of the different species recorded during each sampling session.

At first, the sampling points were grouped into six Nature 2000 habitat types:

- Code 4070 (n = 2; bushes with *Pinus mugo* and *Rhododendron hirsutum*);
- Code 6170 (n = 11; alpine and subalpine calcareous grasslands);
- Code 8120 (n = 17; calcareous and calcshist screes of the montane to alpine levels);
- Code 8210 (n = 12; calcareous rocky slopes with chasmophytic vegetation);

- Code 8240 (n = 3; limestone pavements);
- Code 9420 (n = 10; alpine *Larix decidua* and/or *Pinus cembra* forests) (https://eunis.eea.europa.eu/habitats-annex1-browser.jsp, accessed on 17 January 2021).

As a few sampling points were located in the habitat types 4070 and 8240, we jointed them for affinity (i.e., vegetation cover = 100%), respectively, to the habitat type 9420 and 8210, respectively. Thus, four habitat groups were obtained: 6170; 8120; 8210-8240 and 9420-4070).

The sampling points were then also grouped into seven landform types:

- Active debris flow (n = 2);
- Bedrock (n = 14);
- Glacial deposit (n = 1);
- Large rockslide deposit (n = 10);
- Mature slope (vegetation cover $\geq$ 90%; n = 21);
- Rockslide deposit (n = 1);
- Talus slope (n = 6).

As a few sampling points were located in the active debris flow, glacial deposit and rockslide deposit landform types we combined them for affinity with the talus slope (i.e., small-medium sized rock deposits) and large rockslide deposit landform types, respectively. Consequently, four landform types were considered: bedrock, large rockslide deposit, mature slope and talus slope. Thus, different landform types representing the geomorphological variability that could be found within a habitat type could be examined for each habitat type.

To compare the average species richness, total activity density, taxonomic distinctness and BEP species recorded in each Nature 2000 habitat and landform type we performed a Kruskal–Wallis test. This test was made possible by the fact that the traps set in each Nature 2000 habitat and landform type were spatially independent of each other; thus, the assumption of independence of the observations was satisfied [23,24]. When we obtained a significant difference among habitat types a Dunn's post-hoc test was performed.

To identify characteristic carabid species of each habitat and landform type, indicator species analysis (IndVal) [25] was used. The IndVal index for abundance data was used to quantify the association between each species and each habitat and landform. Once the highest associated habitat and landform were identified for each species, the association was assessed through a permutation test (number of permutations: 999). We then grouped the species in relation to their presence and total activity density for each of the habitat types considered.

To describe the patterns of species distribution in each landform type and their relationships with environmental variables, canonical correspondence analysis (CCA) was used [26]. Before running the analysis, the vegetation cover was arcsin $\sqrt{(p/100)}$ transformed, the altitude was natural log transformed and the aspect was $-\cos(x)$ transformed to normalize the distribution. To run the analysis, the species *Harpalus solitaris* was ruled out from the dataset because it was found in just one sampling point.

All of the analyses were run in PAST 4.05 [26].

3. Results

A total of 928 individuals belonging to 22 species were sampled (Table S1). Half of the sampled species were Alpine endemic and half of them were steno-endemic of the Dolomites (Table S1).

The average species richness and taxonomic distinctness did not show any significant differences among habitat types; conversely, these differences became significant when the landform type was considered (Table 1). The average total activity density and BEP species showed significant differences among both habitat and landform types (Table 1; Table S2; Figure 2).

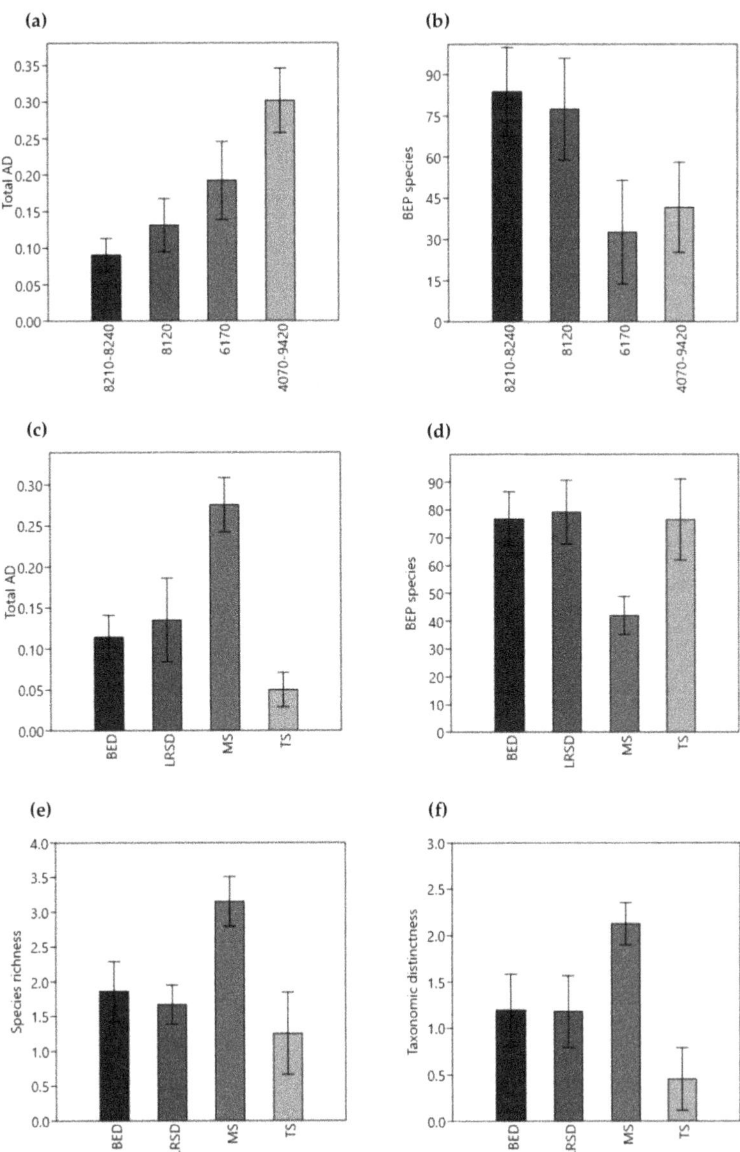

Figure 2. Average carabid species richness, taxonomic distinctness, total activity density and frequency of BEP species (y-axis) in relation to the Nature 2000 habitat (6170 = alpine and subalpine calcareous grasslands; 8120 = calcareous and calcshist screes of the montane to alpine levels; 8210-8240 = limestone pavements and calcareous rocky slopes with chasmophytic vegetation; 9420-4070 = alpine *Larix decidua* and/or *Pinus cembra* forests and bushes with *Pinus mugo* and *Rhododendron hirsutum*); graphs (**a**,**b**) and landform types (BED = bedrock; LRSD = large rockslide deposit; MS = mature slope; TS = talus slope); graphs (**c**–**f**). The whisker intervals represent a 95% confidence interval based on the standard error. Only the results in which we obtained a significant difference ($p < 0.05$) among habitat and landform type are represented (for more information see Table S2).

Table 1. Results of the Kruskal–Wallis test (values in the columns: X^2 and p-value).

Variables	Habitat Type	Landform Type
Species richness	6.3; 0.09	12.3; 0.005 *
Activity density	14.2; 0.003 *	19.2; 0.0002 *
Taxonomic distinctness	5.8; 0.09	8.7; 0.023 *
BEP species	13.4; 0.002 *	10.9; 0.007 *

* significant difference ($p < 0.05$). BEP = brachypterous + endemic + predator.

The most abundant species was *Pterostichus multipunctatus* (sum of AD recorded in all traps in which the carabid was collected = 4.42) followed by *Nebria germari* (sum of AD recorded in all traps in which the carabid was collected = 1.85). The most frequent species resulted *Pterostichus multipunctatus* (47.3% of occurrence) followed by *Carabus adamellicola* (36.4% of occurrence) and *Nebria germari* (23.6% of occurrence); the rest of the species were below 10% occurrence.

According to the indicator species analysis, 12 species were significantly linked to a specific Nature 2000 habitat type: *Nebria germari* was linked to calcareous rocky slopes with chasmophytic vegetation; *Carabus adamellicola* to calcareous scree slopes; *Calathus melanocephalus*, *Carabus convexus*, *Cymindis vaporariorum*, *Harpalus latus* and *H. solitaris* to calcareous grasslands and *Calathus micropterus*, *Cychrus attenuatus*, *Leistus nitidus*, *Pterostichus multipunctatus* and *P. unctulatus* to larch and dwarf pine formations (Table S3).

The graphs for the species assemblages sampled in each Nature 2000 habitat type (Figure 3) show differences in the composition of assemblages found in highly vegetated habitats (code 6170 and 4070-9420) compared with habitats with patchy vegetation and those completely unvegetated (code 8120 and 8210-8240). *Pterostichus multipunctatus* was found to be the species with the highest total activity density (dominant species) exclusively in calcareous grasslands and larch and dwarf pine formations while *Nebria germari* was dominant in calcareous scree slopes with chasmophytic vegetation. All of the habitat types apart from calcareous scree slopes had species assemblages characterized by one dominant species followed by several species with low and similar total activity density values.

According to the indicator species analysis result, 13 species were significantly linked to a specific landform type: *Duvalius nambinensis* and *Oreonebria castanea* to the bedrock landform type; *Carabus adamellicola* and *Carabus creutzeri* to large rockslide deposits; *Calathus melanocephalus*, *C. micropterus*, *Carabus convexus*, *Cychrus attenuatus*, *Cymindis vaporariorum*, *Harpalus latus*, *Pterostichus multipunctatus* and *P. unctulatus* to mature slopes and *Abax pilleri* to talus slopes.

The CCA analysis on species distribution in relation to the Nature 2000 habitat type, landform type and the altitude, vegetation cover and aspect variables revealed that most of the variance was explained by Axis 1 (90.5%) while Axis 2 explained only 9.5% of the total variance (Table 2, Figure 4). As shown in Figure 4, Axis 1 was mainly explained by the negative correlation among altitude and vegetation cover while Axis 2 was weakly explained by aspect. This result was clearly visible in the species distribution along Axis 1; two groups were visible with the former clustered in the quadrants one and two of the CCA and the latter in the quadrants three and four. Specifically, the occurrence of the species *Nebria germari*, *Oreonebria castanea*, *Trechus sinuatus* and *Duvalius nambinensis* was shown as linked to the calcareous scree slopes with chasmophytic vegetation, talus slopes and large rockslide deposits habitats located at the highest altitude and with the lowest vegetation cover. *Carabus creutzeri* and *Carabus adamellicola* were linked to the calcareous scree slopes with chasmophytic vegetation, calcareous grasslands and with bedrock and large rockslide deposit habitats located at an intermediate altitude and with an intermediate level of vegetation cover. The rest of the species (quadrants 3 and 4) were linked to mature slopes and talus slopes located at the lowest altitude and with the highest level of vegetation cover.

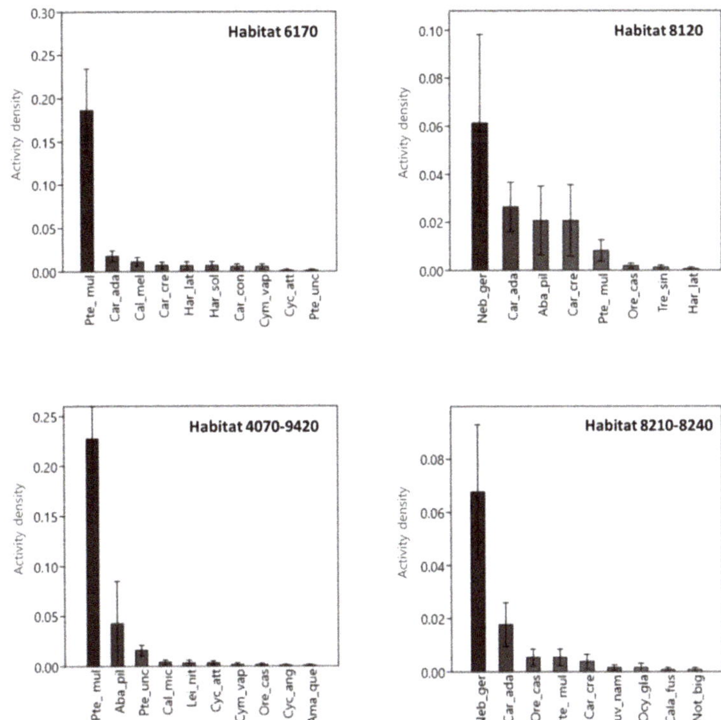

Figure 3. Average species activity density (y-axis) recorded in each of the Nature 2000 habitat types. Habitat types: 6170 = alpine and subalpine calcareous grasslands; 8120 = calcareous and calcshist screes of the montane to alpine levels; 8210-8240 = limestone pavements and calcareous rocky slopes with chasmophytic vegetation; 9420-4070 = alpine *Larix decidua* and/or *Pinus cembra* forests and bushes with *Pinus mugo* and *Rhododendron hirsutum*. The whisker intervals represent a 95% confidence interval based on the standard error. Species abbreviation refers to Table S1.

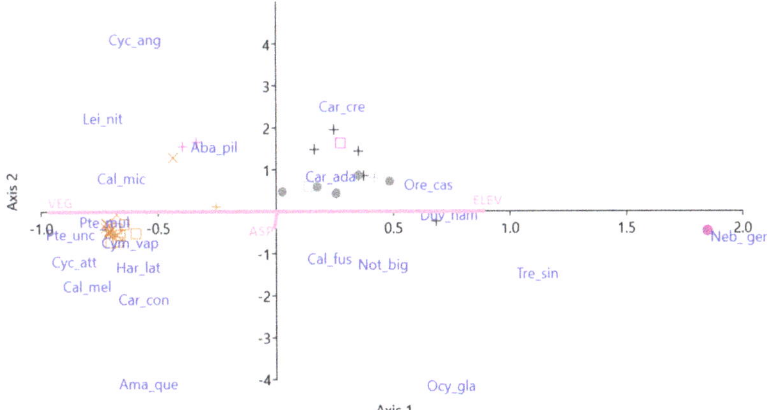

Figure 4. CCA triplot of the distribution of carabid species in relation to the altitude (Alt), vegetation cover (Veg) and aspect (Asp) (vectors in pink). Nature 2000 habitat types are represented as follows: 8210-8240 = black dot; 8120 = plus; 6170 = square; 4070-9420 = X. Landform types are represented as follows: bedrock = dark gray; large rockslide deposit = black; talus slope = fuchsia; mature slope = light brown. Species abbreviation refers to Table S1.

Table 2. Canonical correspondence analysis (CCA) scores of the environmental variables and percentage of explained variance.

CCA	Axis 1	Axis 2
Elevation	0.882	−0.025
Aspect	−0.006	−0.360
Vegetation cover	−0.965	0.001
Eigenvalue	0.912	0.096
% of explained variance	90.5	9.5
p-value (999 permutation)	0.001 *	0.016 *

* significant values, $p < 0.05$.

4. Discussion

Our work was one of the first attempts to simultaneously investigate the effects of habitat and landform types on the distribution of carabid species at a high altitude.

Twenty-two ground-dwelling carabid species were sampled at a high altitude in the Brenta Dolomites.

Interestingly, species richness and taxonomic distinctness changed significantly in relation to landform type but not to habitat type. Specifically, mature slopes had the highest species richness and the highest phylogenetically distant assemblages. The activity density reached the highest values on larch and dwarf pine formations and on mature slopes or on slopes that were stable or least affected by gravitational processes (e.g., rock falls). This finding partially supports the pattern of species richness and activity density observed in the Eastern Dolomites by Pizzolotto et al. [12].

A weak positive correlation between altitude and the frequency of low dispersal (brachypterous), predatory and endemic species was found in the Alps [12]. Our data support the evidence provided by Chamberlain et al. [10] about the role of habitat type as an important driver of the occurrence of species traits. We also demonstrated the role of the landform type in the frequency of BEP species. Specifically, we demonstrated that endemic, poorly-dispersing and predatory species prefer calcareous scree slopes with chasmophytic vegetation and specifically large rockslide deposits, bedrocks and scree slope landforms. The Brenta Dolomites are a peripheral mountain chain of the Italian Alps and compared with the inner massifs [12] they have a higher frequency of species with a low dispersal capability (i.e., steno-endemic species such as *Carabus adamellicola*, *Duvalius nambinensis*) and a specialized diet (e.g., *Carabus creutzeri*, *Cychrus* spp. (snail feeder), *Notiophilus biguttatus* (springtails)).

High altitude habitats of the Brenta Dolomites were dominated by three species: *Nebria germari*, *Carabus adamellicola* and *Pterostichus multipunctatus*. Among them, only *Nebria germari* is also present in the Eastern Dolomites [27]. Currently, *N. germari* does not seem to occur below 2550 m a.s.l.; an altitudinal shift due to the climate warming confirmed throughout the Dolomites [28]. *Carabus adamellicola* is a steno-endemic species of the Adamello-Brenta mountain group, vicariant of *C. bertolinii* and *C. alpestris* of the Central-Eastern Dolomites [12], but unlike these species it was sampled from the alpine calcareous grasslands to substrates with low or patchy vegetation.

The indicator species analysis found that both *Nebria germari* and *Carabus adamellicola* were indicator species of calcareous scree slopes with chasmophytic vegetation. *Nebria germari* was not found to be an indicator of any specific landform within these habitat types. In fact, it can be found in a wider range of landform types such as glacier fronts, moraines along the glacier forelands of the Western Dolomites [19,28] and other fellfields. Conversely, *C. adamellicola* is significantly linked to large rockslide deposits. The presence of these species in high altitude areas can be explained by the local availability of humidity determined by long-lasting snow cover and the microthermal conditions of the fellfield (average annual humidity and temperature of the six plots where the species were found (plot n. 1, 2, 3, 4, 6 and 9, see. Table S1): relative humidity = 92.6 ± 7.8%; average temperature = 2.7 ± 1.4 °C, unpublished data). *Pterostichus multipunctatus*, *Calathus micropterus*,

Cychrus attenuatus, *Leistus nitidus* and *Pterostichus unctulatus* are significantly linked to larch and dwarf pine formations and to the mature slope landform type. Moreover, all of these species except *P. multipunctatus* well represent the typical forest assemblages of the Italian Alps. *Pterostichus multipunctatus* can be considered an ecological vicariant of *P. morio* (Eastern Dolomites) and, like *P. morio* [27], is more common in alpine grasslands located on stable slopes.

The indicator species analysis selected *Calathus melanocephalus*, *Carabus convexus*, *Cymindis vaporariorum*, *Harpalus latus* and *H. solitaris* as species linked to calcareous grasslands. These species are typical of xeric areas [27–29], specifically those that experience the periodical presence of grazing animals (e.g., cows, goats and sheep), which is a quite common practice on the European Alps.

The CCA analysis confirmed the effect of altitude and vegetation cover in driving the species distribution in each habitat and landform type. Specifically, *Nebria germari*, *Trechus sinuatus* and *Duvalius nambinensis* were shown as species linked to talus slopes and large rockslide deposits located at a high altitude (average value = 2600 m a.s.l.) with a low percentage of vegetation cover (average value = 6%).

Carabus creutzeri and *C. adamellicola* distribution was linked to large rockslide deposits and bedrock located at lower altitudes (average value = 2400 m a.s.l.) and with greater vegetation cover (average value = 30%). *Carabus creutzeri* is a forest species [12] but our data support the findings by Brandmayr and Zetto Brandmayr [27] of its presence also in open areas at high altitude, specifically in Alpine areas with a high level of annual rainfall such as the Brenta Dolomites (average annual rainfall = ca. 1400 mm; www.climatrentino.it, accessed on 17 January 2021).

5. Conclusions

Geomorphological diversity (also known as geodiversity) supports a diversity of habitats across a wide range of temporal and spatial scales. It has an important ecological value in supporting biodiversity and ecosystem functioning and it is key element in protected areas [30].

Our data showed that the distribution of carabid beetle species at high altitudes in the Brenta Dolomites area of the Adamello-Brenta Natural Park was driven by two key elements of its mountain slope physiography: habitat and landform type.

The classification approach of the Nature 2000 habitats proved useful for identifying indicator species but geomorphological heterogeneity (landform type) within each habitat type was an additional instrument able to give information about microhabitat preferences. On the basis of the obtained results, we recommend applying a detailed geomorphological approach to studies aimed at investigating the distribution of ground-dwelling arthropods in high altitude areas. We expect that the distribution of other epigeic taxa (e.g., Chilopoda [31]) may also be affected by the geomorphological features of the mountain slopes.

Supplementary Materials: The following are available online at https://www.mdpi.com/article/10.3390/d13040142/s1 Table S1: List of the sampled points with information about the collected species, Elevation (m a.s.l.), Aspect (°), vegetation cover (%), Nature 2000 habitat code and landform type. For each species, information about wing morphology, chorology and diet are provided. Table S2: Dunn's post-hoc values showing the significant differences found for species richness, taxonomic distinctness, activity density and BEP species among the analyzed habitat and landform types. Table S3: Results of the indicator species analysis (IndVal). Significant relationships are reported in bold.

Author Contributions: Conceptualization, A.M., R.C. and M.A.; methodology, M.G.; field activity, A.M.; M.A. and R.C., statistical analysis, M.G.; species sorting and identification, T.B., S.O. and M.G.; writing—original draft preparation, M.G.; project administration, A.M. and V.L.; funding acquisition, A.M., M.G. and V.L. All authors have read and agreed to the published version of the manuscript.

Funding: This research was co-funded by Adamello-Brenta Natural Park and Province Autonomous of Trento.

Institutional Review Board Statement: The permission to collect carabid beetles was provided by the Director of the Adamello Brenta Natural Park (Agreement number: MTSN-0011014-07/12/2018-A—Allegato Utente 1 (A01)).

Informed Consent Statement: Not applicable.

Data Availability Statement: The dataset used in this paper is publicly available in the Supplementary Materials.

Acknowledgments: We are grateful to the staff of the Parco Naturale Adamello-Brenta and to Ivan Petri, for their help during the field activity and to Alessandra Franceschini for her help in managing the carabid beetle collection. David Kavanaugh (California Academy of Science, USA) revised the English and provided useful suggestions.

Conflicts of Interest: The authors declare no conflict of interest.

References

1. Gilardelli, F.; Sgorbati, S.; Armiraglio, S.; Citterio, S.; Gentili, R. Ecological Filtering and Plant Traits Variation across Quarry Geomorphological Surfaces: Implication for Restoration. *Environ. Manag.* **2015**, *55*, 1147–1159. [CrossRef] [PubMed]
2. Gobbi, M. Global warning: Challenges, threats and opportunities for ground beetles (Coleoptera: Carabidae) in high altitude habitats. *Acta Zoöl. Acad. Sci. Hung.* **2020**, *66*, 5–20. [CrossRef]
3. Beron, P. High-mountain Isopoda Oniscidea, Arachnida and Myriapoda in the Old World. Bureschiana, Sofia, Pensoft & Nat. Mus. *Natur. Hist. Sofia* **2008**, *1*, 556.
4. Beron, P. High-altitude: Isopoda, Arachnida and Myriapoda in the New World. Sofia, Pensoft & Nat. Mus. *Natur. Hist. Sofia* **2008**, *2*, 556.
5. Hågvar, S.; Gobbi, M.; Kaufmann, R.; Ingimarsdóttir, M.; Caccianiga, M.; Valle, B.; Pantini, P.; Fanciulli, P.P.; Vater, A. Ecosystem Birth Near Melting Glaciers: A Review on the Pioneer Role of Ground-Dwelling Arthropods. *Insects* **2020**, *11*, 644. [CrossRef] [PubMed]
6. Viterbi, R.; Cerrato, C.; Bassano, B.; Bionda, R.; Von Hardenberg, A.; Provenzale, A.; Bogliani, G. Patterns of biodiversity in the northwestern Italian Alps: A multi-taxa approach. *Community Ecol.* **2013**, *14*, 18–30. [CrossRef]
7. Staunton, K.M.; Nakamura, A.; Burwell, C.J.; Robson, S.K.A.; Williams, S.E. Elevational Distribution of Flightless Ground Beetles in the Tropical Rainforests of North-Eastern Australia. *PLoS ONE* **2016**, *11*, e0155826. [CrossRef] [PubMed]
8. Hiramatsu, S.; Usio, N. Assemblage Characteristics and Habitat Specificity of Carabid Beetles in a Japanese Alpine-Subalpine Zone. *Psyche A J. Èntomol.* **2018**, *2018*, 1–15. [CrossRef]
9. Baranovská, E.; Tajovský, K.; Knapp, M. Changes in the Body Size of Carabid Beetles along Elevational Gradients: A Multispecies Study of Between- and Within-Population Variation. *Environ. Èntomol.* **2019**, *48*, 583–591. [CrossRef]
10. Chamberlain, D.; Gobbi, M.; Negro, M.; Caprio, E.; Palestrini, C.; Pedrotti, L.; Brandmayr, P.; Pizzolotto, R.; Rolando, A. Trait-modulated decline of carabid beetle occurrence along elevational gradients across the European Alps. *J. Biogeogr.* **2020**, *47*, 1030–1040. [CrossRef]
11. Ouisse, T.; Day, E.; Laville, L.; Hendrickx, F.; Convey, P.; Renault, D. Effects of elevational range shift on the morphology and physiology of a carabid beetle invading the sub-Antarctic Kerguelen Islands. *Sci. Rep.* **2020**, *10*, 1–12. [CrossRef] [PubMed]
12. Pizzolotto, R. Habitat diversity analysis along an altitudinal sequence of alpine habitats: The Carabid beetle assemblages as a study model. *Period. Biol.* **2016**, *118*, 241–254. [CrossRef]
13. Bernasconi, M.G.; Borgatti, M.S.; Tognetti, M.; Valle, B.; Caccianiga, M.; Casarotto, C.; Ballarin, F.; Gobbi, M. Checklist ragionata della flora e degli artropodi (Coleoptera: Carabidae e Arachnida: Araneae) dei ghiacciai Centrale e Occidentale del Sorapiss (Dolomiti d'Ampezzo). *Frammenti Conoscere e Tutelare la Natura Bellunese* **2019**, *9*, 49–65.
14. Soldati, M. Dolomites: The Spectacular Landscape of the 'Pale Mountains'. In *Geomorphological Landscapes of the World*; Metzler, J.B., Ed.; Springer: Dordrecht, The Netherlands, 2009; pp. 191–199.
15. Zanoner, T.; Seppi, R.; Carton, A. *Database Geomorfologico del Progetto BioMiti*; Tecnical Report and Vector Data; Adamello Brenta Natural Park: Strembo, Italy, 2019.
16. QGIS Development Team. QGIS Geographic Information System. Available online: http://qgis.osgeo.org,2020 (accessed on 10 February 2021).
17. Clarke, K.R.; Warwick, R.M.; Pienkowski, M.W.; Watkinson, A.R.; Kerby, G. A taxonomic distinctness index and its statistical properties. *J. Appl. Ecol.* **1998**, *35*, 523–531. [CrossRef]
18. Gobbi, M.; Lencioni, V. Do carabids (Coleoptera: Carabidae) and chironomids (Diptera: Chironomidae) exhibit similar diversity and distributional patterns along a spatio-temporal gradient on a glacier foreland? *J. Limnol.* **2018**, *77*, 187–195. [CrossRef]
19. Gobbi, M.; Ballarin, F.; Brambilla, M.; Compostella, C.; Isaia, M.; Losapio, G.; Maffioletti, C.; Seppi, R.; Tampucci, D.; Caccianiga, M. Life in harsh environments: Carabid and spider trait types and functional diversity on a debris-covered glacier and along its foreland. *Ecol. Èntomol.* **2017**, *42*, 838–848. [CrossRef]
20. Pesarini, C.; Monzini, V. Insetti della fauna italiana. Coleotteri Carabidi I. *Società Italiana di Scienze Naturali* **2010**, *100*, 152.
21. Pesarini, C.; Monzini, V. Insetti della fauna italiana. Coleotteri Carabidi II. *Società Italiana di Scienze Naturali* **2011**, *101*, 144.

22. Brandmayr, P.; Brandmayr, T.Z.; Pizzolotto, R. I Coleotteri Carabidi per la valutazione ambientale e la conservazione delle biodiversità. In *Manuale Operativo, Agenzia per la Protezione dell'Ambiente e per i Servizi Tecnici*; IGER: Rome, Italy, 2005; Volume 34, p. 240.
23. Wayne, W.D. Kruskal-Wallis one-way analysis of variance by ranks. In *Applied Nonparametric Statistics*, 2nd ed.; Wayne, W.D., Ed.; Duxbury Classic Series: Boston, MA, USA, 1990; pp. 226–234.
24. Theodorsson-Norheim, E. Kruskal-Wallis test: BASIC computer program to perform nonparametric one-way analysis of variance and multiple comparisons on ranks of several independent samples. *Comput. Methods Programs Biomed.* **1986**, *23*, 57–62. [CrossRef]
25. Dufrene, M.; Legendre, P. Species assemblages and indicator species: The need for a flexible asymmetrical approach. *Ecol. Monogr.* **1997**, *67*, 345–366. [CrossRef]
26. Hammer, Ø.; Harper, D.A.T.; Ryan, P.D. PAST: Paleontological statistics software package for education and data analysis. *Palaeontol. Electron.* **2001**, *4*, 9.
27. Brandmayr, P.; Brandmayr, T.Z. Comunità a coleotteri carabidi delle Dolomiti Sudorientali e delle Prealpi Carniche. Zoocenosi e Paesaggio—I Le Dolomiti, Val di Fiemme-Pale di S. Martino (ed. by P. Brandmayr). *Studi Trent. Sci. Nat. Acta Biol.* **1988**, *64*, 125–250.
28. Valle, B.; Ambrosini, R.; Caccianiga, M.; Gobbi, M. Ecology of the cold-adapted species Nebria germari (Coleoptera: Carabidae): The role of supraglacial stony debris as refugium during the current interglacial period. *Acta Zoöl. Acad. Sci. Hung.* **2020**, *66*, 199–220. [CrossRef]
29. Gobbi, M.; Fontaneto, D.; Bragalanti, N.; Pedrotti, L.; Lencioni, V. Carabid beetle (Coleoptera: Carabidae) richness and functional traits in relation to differently managed grasslands in the Alps. *Annales de la Société entomologique de France (N.S.)* **2015**, *51*, 52–59. [CrossRef]
30. Crofts, R. Linking geoconservation with biodiversity conservation in protected areas. *Int. J. Geoherit. Park.* **2019**, *7*, 211–217. [CrossRef]
31. Gobbi, M.; Caccianiga, M.; Compostella, C.; Zapparoli, M. Centipede assemblages (Chilopoda) in high-altitude landforms of the Central-Eastern Italian Alps: Diversity and abundance. *Rendiconti Lincei Scienze Fisiche e Naturali* **2020**, *31*, 1071–1087. [CrossRef]

Article

Effects of Pure and Mixed Pine and Oak Forest Stands on Carabid Beetles

Alexandra Wehnert *, Sven Wagner and Franka Huth

Department of Forest Sciences, Faculty of Environmental Sciences, Institute of Silviculture and Forest Protection, TU Dresden, 01737 Tharandt, Germany; sven.wagner@tu-dresden.de (S.W.); f.huth@freenet.de (F.H.)
* Correspondence: alexandra.wehnert@freenet.de; Tel.: +49-351-4633-1338

Abstract: The multiple-use approach to forestry applied in Germany aims to combine timber production and habitat management by preserving specific stand structures. We selected four forest stand types comprising (i) pure oak, (ii) equal oak–pine mixtures, (iii) single tree admixtures of oak in pine forest and (iv) pure pine. We analysed the effects of stand composition parameters on species representative of the larger carabid beetles (*Carabus arvensis*, *C. coriaceus*, *C. hortensis*, *C. violaceus*, *Calosoma inquisitor*). The main statistical methods used were correlation analyses and generalised linear mixed models. *Cal. inquisitor* was observed in pure oak forests exclusively. *C. coriaceus* and *C. hortensis* were absent from pure pine stands. High activity densities of *C. arvensis* and *C. violaceus* were observed in all four forest types. When assessed at the smaller scales of species crown cover proportions and spatial tree species effect zones, *C. hortensis* was found to be positively related to oak trees with a regular spatial distribution, whereas *C. coriaceus* preferred lower and more aggregated oak tree proportions. *C. violaceus* showed strong sex-specific tree species affinities. Information about preferences of carabid beetles is necessary for management activities targeting the adaptation of forest structures to habitat requirements.

Keywords: mixed forests; Carabidae; activity density; body size; sex ratio; aggregation index; spatial effect zones

Citation: Wehnert, A.; Wagner, S.; Huth, F. Effects of Pure and Mixed Pine and Oak Forest Stands on Carabid Beetles. *Diversity* **2021**, *13*, 127. https://doi.org/10.3390/d13030127

Academic Editor: Tibor Magura

Received: 26 February 2021
Accepted: 13 March 2021
Published: 17 March 2021

Publisher's Note: MDPI stays neutral with regard to jurisdictional claims in published maps and institutional affiliations.

Copyright: © 2021 by the authors. Licensee MDPI, Basel, Switzerland. This article is an open access article distributed under the terms and conditions of the Creative Commons Attribution (CC BY) license (https://creativecommons.org/licenses/by/4.0/).

1. Introduction

The promotion and protection of mixed forests is considered a high priority in forest landscapes with large areas of artificially established, even-aged coniferous forests [1–3]. A typical example are the single-layered Scots pine (*Pinus sylvestris* L.) forests with naturally occurring admixed sessile oak (*Quercus petraea* (Matt.) Liebl.) trees found in the lowlands of central Europe [4–6]. Mixed forests are characterised by the presence of at least two different tree species [7]. Tree species mixtures can generate positive impacts in terms of the stability and resilience of forest ecosystems [8–10]. The specific traits of tree species, and the stand compositions, increase the potential range of habitat conditions and environmental niches that can be exploited of by different species of flora and fauna [11,12]. The effects of tree species mixtures depend on species proportions, spatial distribution and the patterns of mingling [2,3]. To understand the beneficial effects of mixtures, it is important to identify links between tree species and groups of faunal species [13]. This is all the more important in light of the fact that some studies have shown that greater structural diversity or tree species richness in forest ecosystems is not necessarily associated with an increase in faunal diversity [2,14,15].

Carabid beetles are one of the best investigated groups of beetles and are recognised for their high potential to act as indicator or model species for specific environmental conditions [16–18]. The importance of carabid beetles in forest ecosystems is due to their function as predators and antagonists of pest insects [19,20]. Generalised statements concerning the preferences of carabid beetles in relation to habitat or forest types can lead to misinterpretations or contradictory conclusions, because previous studies of Carabid beetles reported different or unspecific results concerning the beetles' affinities to particular tree species and forest stand types. For example, Lindroth [21] characterised *Carabus violaceus* (L. 1758) as a "eurytopic" species, present in deciduous and coniferous forests. Skłodowski [22] recorded this species in poor coniferous forests only, whereas Dahl [23], by contrast, wrote that *C. violaceus* avoids poor coniferous forests. Another example is given by the following contrasting distributions claimed for *C. hortensis* (L. 1758), variously described as a "dominant species in moderately dry coniferous forests" [24], a species of fertile deciduous forest habitats [22] and a species of sparse deciduous and mixed forests in central Europe [21]. The differing distributions observed for the *Carabus* species can be explained by species-specific behavioural patterns and high locomotory activity [24–26]. Other explanations may derive from the consideration of superordinate spatial levels (e.g., landscape scale), local or regional differences in climate conditions, and the limited assessment of forest stand type characteristics, especially in tree species mixtures [27,28].

More detailed measurements of forest structures will almost certainly increase the value of information concerning the habitat preferences of Carabid beetles [15,29]. One way to increase our knowledge is through an approach to forest inventory that characterises mixed forests on the basis of basic components such as age, dimensions (stem or crown), proportions and the spatial arrangement of the tree species [30,31]. These forest parameters help to characterise forest conditions at smaller spatial scales, such as at the scale of single trees or small groups of trees [32,33]. It has been proven that in the case of single tree admixtures the spatial ecological effects are influenced in particular by individual tree age, diameter at breast height and crown dimension [34,35]. These tree species effects are spatially limited [36] and characterised by distance-dependent gradients or zones [37]. Tree species affect light availability, the microclimate and soil conditions in their surroundings, all of which are relevant for flora and fauna [38,39]. The resultant gradients and zones lead to small-scale edge effects [32]. Consequently, the habitat function of mixed forests as it pertains to Carabid beetles depends on the spatial manifestation of individual tree effects [40]. Soil surface-related tree zone effects are highly relevant for Carabid beetles; for example, humus forms, leaf litter distributions [27,41,42], pH values [39,43,44], topsoil moisture as a result of crown interception [45] and ground vegetation cover [37].

For the purposes of the study of pure and mixed oak and pine forests presented in this paper, we used detailed spatial information tied to oak and pine trees to describe the differences in forest stand types and to analyse related habitat effects for Carabid beetles [27]. The study excluded the effects of different climate regions and temporal aspects such as seasonality of climate conditions or species-specific metamorphic behavior of Carabid beetles [46]. Parameters pertaining to Carabid beetles, such as activity density, body size and sex ratios (number of males to females), can be used to assess the suitability of habitats and their quality [47,48]. The term "activity density" is used in pitfall trap studies (see methods) to record mobile species such as Carabids driven by, for example, intrinsic and extrinsic motivations [25,49]. The activity density represents the conceptual approach behind the pitfall trap method, which aims to provide mobility-related data pertaining to the species present in the context of defined temporal and spatial scales [24,50]. Carabid beetles exhibit strong sex-specific habitat preferences and benefit from the available resources [51,52]. To highlight the relevance of tree species effects within pure and mixed oak and pine forests for the distribution of Carabid beetles, the research was based on the following hypotheses:

1. Differences in activity density, body size and sex ratios can be expected at the level of forest stand types for the five Carabid species analysed (*Carabus arvensis* (Hbst. 1784), *C. coriaceus* (L. 1758), *C. hortensis* (L. 1758), *C. violaceus* (L. 1758), *Calosoma inquisitor* (L. 1758)).
2. Tree species-specific characteristics, such as proportions of crown cover, and the nature of the tree species spatial distribution (random, regular or aggregated) are suitable parameters to highlight the affinities of Carabid beetles, especially for the mixed oak and pine forest stand types.

3. Superordinate analyses are possible, where tree species effect zones (Z1—pure oak effect zone, Z2—mixed oak–pine effect zone, Z3—pure pine effect zone) are used to define small-scale habitat preferences and environmental niches of Carabid species.

2. Materials and Methods

2.1. Study Areas and Tree Species Parameters

All of the sites covered in the study are located to the south of the German Federal State Brandenburg, in the Neusorgefeld forest district (51°47′15.77″ N; 13°34′37.36″ E). The mean distance between the study areas is 1605 m (sd ± 1102 m). The soils typically found in this region of the German lowlands are poor and sandy. The local mean annual temperature during the sampling period in 2011 reached 9.2 °C, and the precipitation amounted to 516 mm [53]. The ground vegetation in the herb and shrub layers of all study sites is dominated by *Deschampsia flexuosa* (L.), *Vaccinium myrtillus* (L.), *Vaccinium vitis-idaea* (L.) and small patches of *Calamagrostis epigejos* ((L.) Roth). *Hypnum* spp. and *Pleurozium schreberi* (Brid.) have the highest abundance in the moss layer. Four different managed forest stand types were considered, characterised by different proportions of the tree species *P. sylvestris* and/or *Q. petraea*: (fst1) pure oak stands, (fst2) even mixtures of oak and pine trees, (fst3) admixtures of less than 10% of oak within a pine forest matrix and (fst4) pure pine stands. Each of the chosen forest stand types included four study areas, except for the third forest stand type, which was represented by three study areas. The total number of study areas was 15, and all study areas were oriented to the north. The tree species distributions of the study areas are shown as position maps in Figure 1, combined with the tree crown dimensions. The position maps are based on polar coordinates measured using a laser-dendrometer (type LEHDA-GEO 100) with a degree of precision of 0.5° for the direction and 0.1 m for the distance. The polar coordinates were transformed into x- and y-coordinates (Cartesian coordinates) for further statistical analyses (see Section 2.3). The study areas established within the forest stand types fst1, fst2 and fst4 were 30 m × 60 m in size. The larger study areas in the forest stand type fst3 covered areas of between 4900 m^2 and 6300 m^2. Diameter at breast height, tree height and crown diameter were recorded for all trees on the 15 study areas. These parameter measurements at the individual tree level were included in further calculations carried out on other spatial scales, for example, at the level of study area (Table 1) and at the level of the areas surrounding traps (e.g., calculation of oak crown cover within 225 m^2), as used in the model approach (see Section 2.3). Stem densities per hectare were calculated per tree species, as were the relative proportions of the tree species and crown covers.

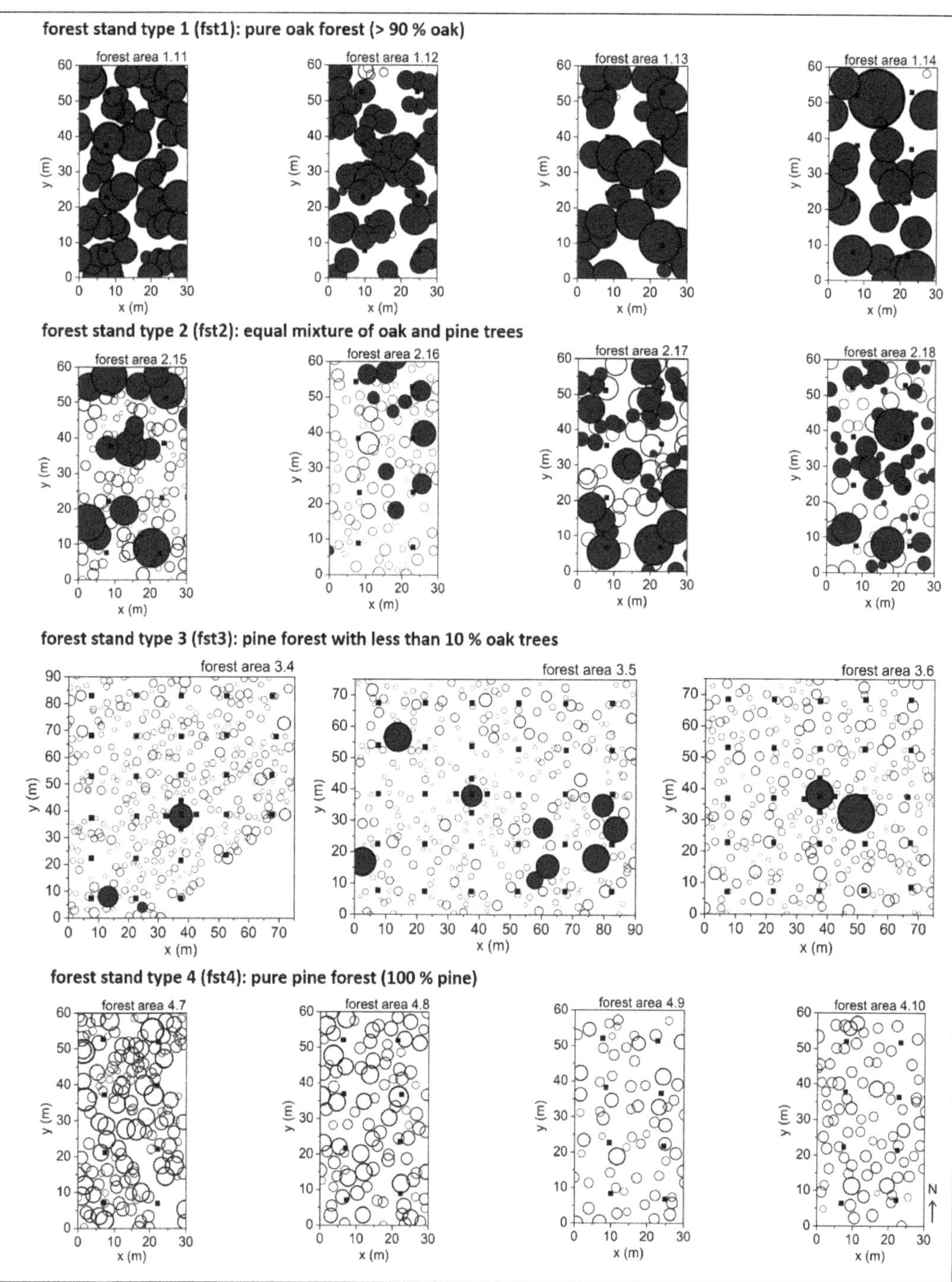

Figure 1. Maps of tree and pitfall trap positions maps (white circles—pine trees, dark grey circles—oak trees, black squares—pitfall traps).

Table 1. Forest stand parameters characterising the study sites representing the four different forest types (fst1—pure oak, fst2—mixed oak-pine, fst3—<10% oak, fst4—pure pine), namely age (years), number of trees per ha (n/ha) and species, proportion of trees per species, relative crown cover (%) per tree species, mean diameter at breast height—dbh (cm), mean height (m), mean crown diameter (m) and aggregation index (R) according to Clark and Evans (R = 1 random distribution, R < 1 clustered/aggregated distribution, R > 1 regular/even distribution) with CSR test, p-value ≤ 0.05 *).

	Forest Area		No. 1.11		No. 1.12		No. 1.13		No. 1.14			No. 2.15		No. 2.16		No. 2.17		No. 2.18	
Parameters	**Units**		**Pine**	**Oak**	**Pine**	**Oak**	**Pine**	**Oak**	**Pine**	**Oak**		**Pine**	**Oak**	**Pine**	**Oak**	**Pine**	**Oak**	**Pine**	**Oak**
age	(years)			122	57	111	37	144	49	197		62	94	99	59	139	104	81	92
no. of trees	(n/ha)			372	28	371	11	167	11	122		600	106	483	67	206	217	194	256
prop. of trees	(%)			100	7	93	6	94	8	92		85	15	88	12	49	51	43	57
crown cover	(%)			62.4	1.0	53.9	0.2	61.7	0.2	57.4		20.0	30.5	22.1	10.9	18.6	41.0	14.3	36.8
dbh (mean)	(cm)	**fst 1**		30.92	16.44	28.25	17.10	47.26	24.08	49.93	**fst 2**	23.48	31.50	25.68	29.78	38.86	19.34	32.55	23.21
height (mean)	(m)			17.67	14.21	16.55	10.80	21.76	17.70	20.29		19.37	19.35	19.78	17.99	24.40	14.53	22.67	19.16
crown diameter (mean)	(m)			6.05	3.62	5.12	3.59	8.49	3.27	8.41		3.69	6.53	4.22	6.36	5.04	5.89	4.66	4.75
aggregation index (R)				1.095	0.877	1.010	0.450	1.146	0.856	1.295 *		1.190 *	0.925	1.281 *	1.147	1.238 *	1.079	1.359 *	1.008

Forest Area			No. 3.4		No. 3.5		No. 3.6			No. 4.7	No. 4.8	No. 4.9	No. 4.10
Parameters	**Units**		**Pine**	**Oak**	**Pine**	**Oak**	**Pine**	**Oak**		**Pine**	**Pine**	**Pine**	**Pine**
age	(years)		53	150	57	150	60	150		57	60	87	104
no. of trees	(n/ha)		645	5	547	13	526	5		817	544	433	456
prop. of trees	(%)		99	1	98	2	99	1		100	100	100	100
crown cover	(%)		24.0	1.8	19.3	5.8	24.1	3.6		61.5	60.9	26.4	34.5
dbh (mean)	(cm)	**fst 3**	24.05	38.02	24.36	44.03	25.69	48.32	**fst 4**	24.09	24.71	30.41	29.74
height (mean)	(m)		20.45	16.07	20.12	18.66	21.18	17.87		21.58	19.63	20.93	22.71
crown diameter (mean)	(m)		2.97	8.13	2.84	10.33	3.34	12.84		3.10	3.15	2.35	2.65
aggregation index (R)			1.116 *	0.635	1.196 *	1.029	1.249 *	0.220		1.158 *	1.360 *	1.225 *	1.316 *

2.2. Trap Design and Beetle Parameters

The positions of the pitfall traps were recorded in the same way as the tree positions (Figure 1). A trap arrangement in a 15 m × 15 m grid was employed on all study areas, according to the body size of the Carabidae [54]. The use of pitfall traps is a common method to record ground-dwelling arthropods [50,55]. Eight pitfall traps were placed at each sampling site for the forest types fst1, fst2 and fst4 with an area of 1800 m^2. In the case of the lager areas (4900 m^2 and 6300 m^2) for forest stand type fst3, the same grid was used for the pitfall traps, with four additional traps placed around one large oak tree. This produced trap numbers of between 29 and 34 per forest type (Table 2). The pitfall traps, made of glass, were 7.5 cm in diameter and 9 cm in depth. They were filled to 75% with a solution of saturated benzoic acid and detergent [56] and covered by a transparent plastic roof to exclude precipitation. The use of 5% benzoic acid has proven successful as a killing liquid and preservative to record Carabids within forest ecosystems. The solution of saturated benzoic acid has the following positive properties compared to other killing liquids and preservatives: (i) no toxicity, (ii) the preservative effect prevents disintegration by fungi or putrid bacteria, (iii) no discolouration, (iv) no odour development and (v) no hardening of recorded Carabids [50,57–59]. The control interval for the 190 pitfall traps was 14 days, synchronised for all study areas. The related activity densities of the Carabid beetles were calculated as a number of specimens per m^2. The control period for the pitfall traps extended from 16 March until 27 October 2011.

Table 2. Number of traps per zone (i) and proportions of zones (ii) per study site and specific to the particular forest stand type (fst).

Forest Stand Type		Stand No.	No. of Traps per Zone			Proportion of Zones (%)		
			Z1 (oak)	Z2 (oak-pine)	Z3 (pine)	Z1 (oak)	Z2 (oak-pine)	Z3 (pine)
fst1	pure oak	1.11	8	0	0	100	0	0
		1.12	8	0	0	100	0	0
		1.13	8	0	0	100	0	0
		1.14	8	0	0	100	0	0
fst2	oak–pine mix.	2.15	3	5	0	34	66	0
		2.16	3	5	0	26	68	6
		2.17	3	5	0	30	69	2
		2.18	2	6	0	26	74	0
fst 3	pine with <10% oak	3.4	2	6	23	2	14	84
		3.5	1	19	14	6	48	45
		3.6	1	8	20	2	15	83
fst 4	pure pine	4.7	0	0	8	0	0	100
		4.8	0	0	8	0	0	100
		4.9	0	0	8	0	0	100
		4.10	0	0	8	0	0	100

The spatial relations were established using the trap coordinates. The body length (cm), height (cm) and width (cm) of all sampled Carabid beetles (imagines) were measured to calculate the sex-specific (females versus males) individual body size (volume in cm^3) by means of the ellipsoid formula [60]. The sex ratio for all Carabid species was determined by dividing the number of males by the number of females [61].

2.3. Aggregation Index and Statistical Analyses

The calculation of aggregation indices based on the method of nearest-neighbour distances is one possibility to get additional information about the spatial distribution of tree species, in particular within mixed forests [62,63]. We used the Clark–Evans aggregation index (R) including the "Donnelly" edge correction and a Monte Carlo test based on 999 simulations of CSR with fixed n [64]. The null hypothesis states a completely random spatial distribution of trees, whereas the alternative hypothesis describes a clustered or regular distribution pattern. The values of the aggregation index R can be interpreted as follows: R = 1 indicates completely random distribution, R > 1 regular or even distribution and R < 1 aggregated or clustered distribution.

Following a stepwise adaptation of the data analyses to the different hierarchies of spatial scale, we started with the forest stand type level, followed by individual forest areas and finally small-scale tree effect zones. At the level of forest stand type, analyses of variance (ANOVA) and additional multiple post-hoc tests (LSD, least significant difference) were used to test for differences in the mean species- and sex-specific densities, body sizes and sex ratios of Carabid beetles between the four different forest stand types (Figure 2). Bivariate correlation analyses (Pearson's correlation coefficient, p-value ≤ 0.05) were carried out to test the strength of the relationships between the tree species-specific parameters (crown cover proportions and aggregation indices) and the species-specific parameters (density and body size) of the Carabid beetles at the forest stand level. The value of the correlation coefficient (r) ranged between +1 and -1 [65]. Distance-dependent single tree effect zones were calculated for each forest area for the small-scale analyses [66,67]. The following three spatial effect zones were defined according to Wehnert and Wagner [42]: Z1 describes the distance between the oak trunk and the crown edge projection (0 m to < 4 m), Z2 is the distance between ≥ 4 m to < 15 m characterised by mixed oak–pine conditions and Z3 is the distance ≥ 15 m representative for pure pine parts of the stands in the study areas. Table 2 gives an overview of numbers and proportions of zones for each study area. Unexpectedly, in forest stand type fst2, with a roughly even mixture of oak and pine trees, the resultant proportions of pure pine zones (Z3) were low. The opposite is true for the study areas of the forest stand type fst3 with less than 10% oak trees.

The detailed information obtained at the small-scale was integrated employing a generalised linear mixed model (GLMM) approach [68]. The use of GLMMs allowed for the inclusion of the spatially explicit small-scale data collated gathered in the immediate surroundings of the pitfall traps, such as the effect zones and the canopy cover proportions of the tree species differentiated by the defined forest stand types [69]. The model adopted "forest stand type" and "tree effect zone" or "proportion of oak crowns" as fixed effects and "forest area" as a random effect to explain the species- and sex-specific densities of Carabid beetles. GLMMs were fitted by means of a negative-binomial error distribution and a logarithmic link. The applied GLMM structure was:

$$rv_{i,k} = \beta_0 + \beta_1 forest.stand.type_i + \beta_2 oak.crown.cover_i + \mu_k + \varepsilon \tag{1}$$

taken the oak effect as a metric covariate or

$$rv_{i,k} = \beta_0 + \beta_1 forest.stand.type_i + \beta_2 forest.zone_i + \mu_k + \varepsilon \tag{2}$$

taken the oak effect as a categorical covariate with

$$forest.zone_i = \begin{cases} 0, & oak\ distance < 4\ \text{m} \\ 1, & 4 \leq oak\ distance < 15\ \text{m} \\ 2, & oak\ distance \geq 15\ \text{m} \end{cases} \tag{3}$$

with rv_i as the response variable (i.e., species- or sex-specific number of beetles) of the ith trap and μ_k as the random effect parameter of the kth stand. We tested model residuals (qq-plots), heteroscedasticity in the residuals, and we checked for spatial autocorrelation by semi-variograms. Therefore, it was not necessary to adjust the original model.

The GLMM outputs were used for predictions of species- and sex-specific beetle activity densities differentiated by forest stand types and zones (Figure 3). The R package "lmmTMB" was used in combination with the package "lme4" for the calculations [70]. The models were fitted using maximum likelihood estimation via "TMB" (Template Model Builder). All calculations were conducted using R software version 4.0.3 (10 October 2020).

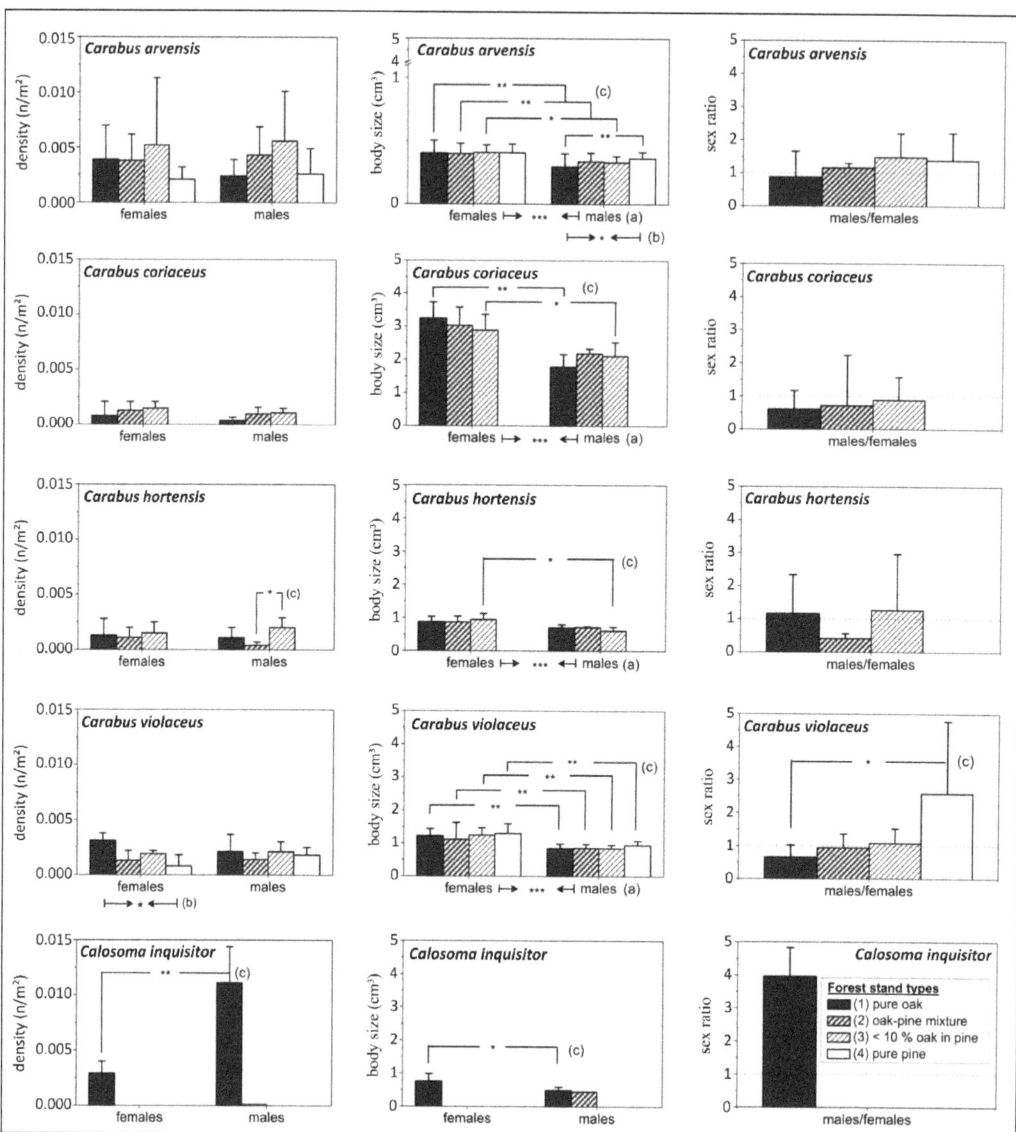

Figure 2. Overview of the parameters for *C. arvensis*, *C. coriaceus*, *C. hortensis*, *C. violaceus* and *Cal. inquisitor* (left column: densities (n/m^2), central column: body size (cm^3), right column: sex ratio (males to females)) related to the four different forest stand types and including the results of the following statistical tests: (**a**) ANOVA for males versus females, all forest stand types included, (**b**) separate ANOVA for females and males across all forest stand types, and (c) multiple post-hoc tests (LSD) across all forest stand types and for both sexes (The levels of significance: '***' 0.001, '**' 0.01, '*' 0.05).

3. Results

3.1. Effects of Forest Stand Types, Tree Species Proportions and Effect Zones

The total numbers of adult specimens recorded per Carabid species over the whole sampling period of 7.5 months and across all of the study areas were as follows: 337 *C. arvensis*, 70 *C. coriaceus*, 75 *C. hortensis*, 144 *C. violaceus* and 102 *Cal. inquisitor*. *C. arvensis* and *C. violaceus* were present in all four forest stand types. *C. arvensis* exhibited the highest activity densities of individuals across all forest stand types. *Cal. inquisitor* was also present at high activity densities but occurred only in the pure oak forest stand type (fst1). The activity densities of males of *Cal. inquisitor* were significantly higher than of females (Figure 2). *C. coriaceus* and *C. hortensis* were completely absent from pure pine stands (fst4). The body sizes differed between the sexes of all Carabid species, females generally being larger.

For both sexes of *C. arvensis* the highest activity densities were observed in the forest stand type with <10% oak trees (fst3). The activity densities of females were higher in forest stand types where oaks were present, while *C. arvensis* males responded to higher proportions of pine trees with higher activity densities. Forest stand type had no effect on the body size of *C. arvensis* females. However, an increase in the proportions of pine in the different forest stand types was associated with a slight increase in the size of *C. arvensis* males. Indeed, a significant difference (p-value = 0.01) in male body size was observed between the pure oak and pure pine stands. The sex ratio increased to values above 1, representing the dominance of males, with an increase in the pine tree proportions across the forest stand types.

The activity densities of *C. coriaceus* were comparatively low but exhibited a positive trend towards the forest stand type with <10% of oak. The activity densities were low in pure oak, whereas individuals of both sexes were missing in pure pine stands. The forest stand types revealed sex-specific effects on *C. coriaceus* body sizes. The body size of the females was largest in pure oak stands. Males were largest in the oak–pine mixed stands (fst2). The sex ratio of *C. coriaceus* approached a balanced value (approx. 1) with an increasing proportion of pines in the forest stand types.

The activity density of *C. hortensis* was comparable to that of *C. coriaceus*. Females and males (imagines) of *C. hortensis* were missing in pure pine stands. The activity densities of females were similar across the studied forest stand types. The activity densities of males differed significantly between the mixed oak–pine stand type and the stand type with <10% oak (p-value = 0.05). The sexes of *C. hortensis* exhibited comparable body sizes across the different forest stand types, except the forest stand type with <10% oak. In the latter case, a significant difference between the body sizes of the sexes was observed (p-value = 0.05). In the mixed oak–pine forest stand type females predominated (sex ratio < 1), but the sex ratio of the other forest stand types highlighted a predominance of males.

The activity densities of *C. violaceus* females differed significantly between the forest stand types (p-value = 0.05). For both sexes the highest activity densities occurred in the forest stand types with pure oak (fst1) and with <10% oak (fst3). The sex-specific body sizes were comparable for each of the different forest stand types. Significant differences in body sizes between the sexes were observed for each forest stand type (p-value = 0.01). In forest stand types with higher proportions of pine the predominance of males increased continuously up to ratios of above 1. *C. violaceus* females were predominant in the pure oak stands (fst1). This resulted in significant differences in the sex ratios between the pure oak and pure pine forest stands (p-value = 0.05).

Cal. inquisitor was only observed in pure oak stands, except for one male that was recorded in the oak–pine mixed forest stand type. The significant differences (p-value = 0.01) between the sexes were reflected by a sex ratio of 3.97, which is representative of a strongly male-dominated distribution.

The relationship (Table 3) between oak crown cover proportions and the activity density of females was positively correlated for *C. hortensis* (r = 0.508, p-value = 0.05), *C. violaceus* (r = 0.676, p = 0.01) and *Cal. inquisitor* (r = 0.814, p = 0.001). This correlation was also positive for *C. coriaceus* (r = 0.236, n.s.) but not significantly so. A strongly positive correlation between the crown cover of oaks and the activity density of both sexes was determined only for *Cal. inquisitor* (r = 0.811, p-value = 0.001). All female activity densities (except *C. arvensis*) were significantly negatively correlated with the proportion of pine crown cover.

Table 3. Correlations (Pearson's correlation coefficient r) between the proportion of species-specific canopy cover (oak and pine trees) and beetle activity density and body size. (The levels of significance: '***' 0.001, '**' 0.01, '*' 0.05, '.' 0.1; significant results are indicated by bold font).

Forest Stand Parameters	Carabid Parameters	*Carabid Species*	Oaks						Pines					
			Females	*p*	Males	*p*	Imagines	*p*	Females	*p*	Males	*p*	Imagines	*p*
crown cover (%)	beetle density	*C. arvensis*	0.018	n.s.	−0.277	n.s.	−0.122	n.s.	−0.179	n.s.	0.098	n.s.	−0.046	n.s.
		C. coriaceus	0.294	n.s.	0.407	n.s.	0.236	n.s.	**−0.412**	.	**−0.466**	.	**−0.420**	.
		C. hortensis	**0.508**	*	0.259	n.s.	**0.462**	.	**−0.516**	*	−0.394	n.s.	**−0.548**	*
		C. violaceus	**0.676**	**	−0.054	n.s.	**0.485**	.	**−0.526**	*	0.191	n.s.	**−0.441**	.
		Cal. inquisitor	**0.814**	***	**0.806**	***	**0.811**	***	**−0.668**	**	**−0.670**	**	**−0.670**	**
	body size	*C. arvensis*	0.165	n.s.	**−0.649**	**	−0.335	n.s.	−0.340	n.s.	**0.492**	.	0.100	n.s.
		C. coriaceus	0.080	n.s.	0.330	n.s.	0.204	n.s.	−0.360	n.s.	**−0.487**	.	**−0.466**	.
		C. hortensis	**0.446**	.	**0.560**	*	**0.566**	*	**−0.552**	*	**−0.565**	*	**−0.636**	**
		C. violaceus	0.286	n.s.	**−0.479**	.	0.034	n.s.	−0.132	n.s.	**0.501**	*	0.207	n.s.
		Cal. inquisitor	**0.872**	***	**0.830**	***	**0.880**	***	NA	-	NA	-	NA	-
aggregation index trees (R)	beetle density	*C. arvensis*	0.231	n.s.	0.132	n.s.	0.193	n.s.	−0.291	n.s.	0.135	n.s.	−0.060	n.s.
		C. coriaceus	**0.435**	.	**0.540**	*	**0.454**	.	**−0.436**	.	−0.287	n.s.	−0.268	n.s.
		C. hortensis	**0.610**	.	**0.468**	.	**0.671**	**	**−0.510**	.	−0.380	n.s.	**−0.506**	*
		C. violaceus	0.423	n.s.	−0.145	n.s.	0.274	n.s.	**0.507**	*	**0.492**	*	**0.497**	*
		Cal. inquisitor	0.436	n.s.	0.396	n.s.	0.406	n.s.	**−0.782**	**	**−0.737**	*	**−0.752**	**
	body size	*C. arvensis*	0.189	n.s.	−0.159	n.s.	0.007	n.s.	−0.217	n.s.	**0.811**	***	**0.452**	.
		C. coriaceus	0.293	n.s.	−0.105	n.s.	0.118	n.s.	−0.415	n.s.	**0.784**	*	0.297	n.s.
		C. hortensis	−0.003	n.s.	0.424	n.s.	−0.007	n.s.	0.264	n.s.	0.130	n.s.	0.130	n.s.
		C. violaceus	0.258	n.s.	−0.119	n.s.	−0.039	n.s.	−0.114	n.s.	0.387	n.s.	0.155	n.s.
		Cal. inquisitor	**−0.961**	*	0.749	n.s.	−0.059	n.s.	NA	-	NA	-	NA	-

Very strong positive relationships existed between oak crown cover proportions and body sizes of *Cal. inquisitor* (r = 0.880, *p*-value = 0.001). The positive correlations between oak crown proportions and the body size of Carabids were also identified as being significant for *C. hortensis* (r = 0.566, *p*-value = 0.05), but this relationship was significantly negative with regard to the crown proportion of pine trees (r = −0.636, *p*-value = 0.01). Negative tendencies for the relationships between the pine crown proportion and the body size of beetles were determined for *C. coriaceus* imagines (r = −0.466, *p*-value = 0.1). Positive effects of the pine crown cover proportion were identified for the body size of males of *C. arvensis* (r = 0.492, *p*-value = 0.1) and *C. violaceus* (r = 0.501, *p*-value = 0.05). The effect of the proportion of pine crown cover was negative for the body size of females of both species.

The correlation coefficients between the aggregation indices of tree species and parameters of Carabid beetles (Table 3) revealed positive effects of more regular oak tree distributions on the activity densities of *C. coriaceus* (r = 0.454, *p*-value = 0.1) and *C. hortensis* (r = 0.671, *p*-value = 0.01). Greater regularity of spatial pine tree distribution caused significantly higher activity densities of both sexes of *C. violaceus* (*p*-value = 0.05). The negative correlation coefficients indicated higher activity densities, especially of females, of *C. coriaceus* (r = −0.436, *p*-value = 0.1), *C. hortensis* (r = −0.510, *p*-value = 0.1) and *Cal. inquisitor* (r = −0.782, *p*-value = 0.01) where the pine trees were more aggregated. The spatial distributions of tree species were less related to body sizes of Carabid beetles (Table 3).

Only the body size of *C. arvensis* (r = 0.811, *p*-value = 0.001) and *C. coriaceus* (r = 0.784, *p*-value = 0.05) males were positively correlated with the aggregation index of pine trees, which means that a more regular distribution of pine trees was linked to an increase in body sizes. The negative correlation between the oak tree aggregation index and the body sizes of *Cal. inquisitor* females was representative of a positive effect on body size of a greater aggregation of oak trees.

Table 4 shows additional information on the smaller spatial scale represented by the effect zones and their effects on the beetle species and sexes. The activity densities of both sexes of *Cal. inquisitor* (r = 0.920, *p*-value = 0.001) and of females of *C. violaceus* (r = 0.654, *p*-value = 0.05) were positively correlated with the proportion of the oak zone (Z1). The density of *C. coriaceus* imagines showed the strongest positive correlation (r = 0.502, *p*-value = 0.1) with the mixed oak–pine zone (Z2). A tendency towards a positive correlation was also observed for *C. hortensis* imagines (r = 0.273, n.s.) and the males of *C. arvensis* (r = 0.124, n.s.). Significantly negative correlations with Z3 were determined for *C. hortensis* females (r = −0.520, *p*-value = 0.05) and imagines (r = −0.507, *p*-value = 0.05) and males of *C. coriaceus* (r = −0.490, *p*-value = 0.1).

Positive correlation coefficients were determined for the body sizes of *C. coriaceus* females (r = 0.541, *p*-value = 0.1) and both sexes of *C. hortensis* (r = 0.537, *p*-value = 0.05) within the pure oak zone. Positive correlations were also found between the body sizes of males of both *C. arvensis* (r = 0.575, *p*-value = 0.05) and *C. coriaceus* (r = 0.593, *p*-value = 0.1) with an increasing proportion of pine zone (Z3), whereas the sizes of males of *C. hortensis* were negatively correlated with increasing pine zone (r = −0.630, *p*-value = 0.1).

Table 4. Correlations (Pearson's correlation coefficient r) between proportions of zones (Z1—oak, Z2—oak–pine mixture, Z3—pine) and activity densities and body sizes of carabid beetles. (The levels of significance: '***' 0.001, '**' 0.01, '*' 0.05, '.' 0.1; significant results are indicated by bold font).

	Carabid Species	Oak Zone (Z1)						Oak–Pine Zone (Z2)						Pine Zone (Z3)					
		Fem	*p*	Mal	*p*	Imag	*p*	Fem	*p*	Mal	*p*	Imag	*p*	Fem	*p*	Mal	*p*	Imag	*p*
beetle density	*C. arv.*	0.051	n.s.	−0.250	n.s.	−0.099	n.s.	0.009	n.s.	0.215	n.s.	0.124	n.s.	−0.052	n.s.	0.084	n.s.	0.007	n.s.
	C. cor.	0.180	n.s.	0.214	n.s.	0.074	n.s.	0.341	n.s.	**0.443**	.	**0.501**	.	−0.390	n.s.	**−0.490**	.	−0.401	n.s.
	C. hort.	0.375	n.s.	0.220	n.s.	0.358	n.s.	0.269	n.s.	0.105	n.s.	0.273	n.s.	**−0.520**	*	−0.269	n.s.	**−0.507**	*
	C. viol.	**0.654**	*	0.004	n.s.	**0.497**	*	−0.277	n.s.	−0.305	n.s.	−0.339	n.s.	−0.408	n.s.	0.201	n.s.	−0.225	n.s.
	Cal. in.	**0.925**	***	**0.915**	***	**0.920**	***	NA	-	NA	-	NA	-	NA	-	NA	-	NA	-
body size	*C. arv.*	0.254	n.s.	**−0.639**	**	−0.265	n.s.	−0.100	n.s.	0.007	n.s.	−0.065	n.s.	−0.164	n.s.	**0.575**	*	0.284	n.s.
	C. cor.	**0.541**	.	0.366	n.s.	0.475	n.s.	−0.077	n.s.	**0.593**	.	0.404	n.s.	−0.240	n.s.	**0.593**	.	0.023	n.s.
	C. hort.	**0.502**	.	**0.631**	*	**0.537**	*	−0.004	n.s.	0.041	n.s.	−0.096	n.s.	−0.002	n.s.	**−0.630**	.	−0.035	n.s.
	C. viol.	0.196	n.s.	−0.352	n.s.	−0.059	n.s.	0.269	n.s.	−0.142	n.s.	−0.012	n.s.	−0.088	n.s.	0.414	n.s.	0.062	n.s.
	Cal. in.	NA	-	0.688	n.s.	**0.819**	.	NA	-	NA	-	NA	-	NA	-	NA	-	NA	-

3.2. Model-Based Effects of Forest Stand Types and Different Structural Attributes

3.2.1. Modelling the Effects of Forest Stand Types and Oak Crown Cover

The first generalised linear mixed model combined the effect of forest stand types and the proportion of oak crown cover (see formula 1). *Cal. inquisitor* was excluded from the following analyses, because this species was only present in the forest stand type with pure oak. This first model (Table 5) revealed no significant effects on the activity density of *C. arvensis*. For *C. violaceus* a significant effect (p-value ≤ 0.05) of the mixed oak–pine stand type (fst2) was only estimated by the model.

Table 5. Parameters for the number of carabid beetles (*C. arvensis*, *C. coriaceus*, *C. hortensis*, *C. violaceus*) using the generalised linear mixed model (GLMM) differentiated by the forest stand types (fst 1—pure oak, fst 2—oak-pine mixture, fst 3—<10% oak, fst 4—pure pine) and combined with the proportion of oak crown cover (cc in%) around each trap position. The levels of significance are indicated by: '***' 0.001, '**' 0.01, '*' 0.05, '.' 0.1.

No. of Beetles		Females				Males				Imagines			
		Estimate	**std. Error**	***p*-Value**		**Estimate**	**std. Error**	***p*-Value**		**Estimate**	**std. Error**	***p*-Value**	
C. arvensis	interc.	−0.3486	0.6861	0.611		−0.9919	0.6883	0.150		−0.0281	0.5520	0.959	
	fst 2	−0.0135	0.6499	0.983		0.8178	0.5915	0.167		0.4309	0.5255	0.421	
	fst 3	−0.0336	0.7484	0.964		0.8762	0.6981	0.209		0.4549	0.6162	0.460	
	fst 4	−0.5006	0.8035	0.533		0.3924	0.7666	0.609		0.0271	0.6484	0.967	
	oak cc	0.0006	0.0079	0.944		0.0036	0.0079	0.651		0.0031	0.0060	0.608	
C. coriaceus	interc.	−3.2320	0.9403	0.001	***	−4.7730	1.0460	0.000	***	−3.0466	0.7934	0.000	***
	fst 2	1.2420	0.7370	0.092	.	1.7040	0.8671	0.049	*	1.4339	0.6162	0.020	*
	fst 3	1.5810	0.8680	0.068	.	3.0240	0.9322	0.002	**	2.1183	0.7173	0.003	**
	fst 4	−18.7500	10,760.00	0.999		−17.6800	1383.00	0.999		−18.4916	9404.59	0.999	
	oak cc	0.0214	0.0114	0.060	.	0.0344	0.0114	0.003	**	0.0243	0.0092	0.008	**
C. hortensis	interc.	−5.4180	1.3420	0.000	***	−3.8935	1.5625	0.013	*	−3.9899	1.1684	0.001	***
	fst 2	1.9110	0.9020	0.034	*	0.5937	1.0665	0.578		1.3733	0.7862	0.081	.
	fst 3	3.0090	1.1146	0.009	**	1.8370	1.3474	0.173		2.4921	1.2026	0.015	*
	fst 4	−16.6500	10,720.00	0.999		−17.6840	9896.70	0.999		−17.5303	9620.49	0.999	
	oak cc	0.0548	0.0151	0.000	***	0.0359	0.0185	0.053	.	0.0461	0.0134	0.001	***
C. violaceus	interc.	−0.6633	0.6856	0.333		0.2344	0.7949	0.768		0.3961	0.5529	0.474	
	fst 2	−0.7316	0.5462	0.180		−0.9195	0.6441	0.153		−0.9192	0.4485	0.040	*
	fst 3	−0.3481	0.6424	0.588		−1.1208	0.7760	0.149		−0.6468	0.5300	0.220	
	fst 4	−0.9944	0.8048	0.217		−1.1232	0.8556	0.189		−0.8987	0.6131	0.143	
	oak cc	0.0039	0.0092	0.668		−0.0515	0.0109	0.169		−0.0040	0.0075	0.592	

Forest stand type effects linked to the proportions of oak crown cover were identified for the presence of *C. coriaceus* and *C. hortensis*. The results of this GLMM revealed a significant effect of oak crown cover on the imagines of *C. coriaceus* (p-value ≤ 0.01) and *C. hortensis* (p-value ≤ 0.001). Sex-specific differences were evident. For example, the forest stand types with oak tree admixtures and higher oak crown cover proportions had a more significant effect on females of *C. hortensis* than on males. The opposite was observed for *C. coriaceus*, where the effect of forest stand type was greater for males, in particular in the pine forests with less than 10% oak (p-value = 0.01).

3.2.2. Modelling the Effects of Forest Stand Types and Effect Zones

The second GLMM (see formula 2) included the four different forest stand types and the ecological effect zones (Z1 to Z3). The GLMM components again provided no significant explanation for the number of *C. arvensis* individuals (Table 6). However, this model better explained the number of *C. violaceus* imagines, which were influenced significantly by the different forest stand types. The combination of forest stand type and tree species-related effect zones resulted once again in significant model estimations for *C. coriaceus* and *C. hortensis*. The tree effect zones were determined to be highly significant in terms of estimating the number of imagines and males of *C. coriaceus* and *C. hortensis* (p-value ≤ 0.001). The mixed forest stand type (fst2), with roughly equal proportions of oak and pine, revealed lower (C. *coriaceus*) to no (C. *hortensis*) relevance as a means to predict the number of beetles.

Table 6. Parameters for the number of carabid beetles (*C. arvensis, C. coriaceus, C. hortensis, C. violaceus*) using the generalised linear mixed model (GLMM) differentiated by the forest stand types (fst 1—pure oak, fst 2—oak–pine mixture, fst 3—less than 10% oak, fst 4—pure pine) and combined with the three tree species-related effect zones (Z1—oak, Z2—oak–pine mixture, Z3—pine). The levels of significance are indicated by: '***' 0.001, '**' 0.01, '*' 0.05, '.' 0.1.

No. of Beetles		Females				Males				Imagines			
		Estimate	Std. Error	*p*-Value		Estimate	Std. Error	*p*-Value		Estimate	Std. Error	*p*-Value	
C. arvensis	interc.	−0.3108	0.3914	0.427		−0.7406	0.3767	0.049	*	0.1922	0.3400	0.572	
	fst 2	−0.3304	0.6209	0.595		0.4308	0.5528	0.436		0.0119	0.5114	0.981	
	fst 3	−0.4359	0.6658	0.513		0.5275	0.5902	0.371		−0.0104	0.5490	0.985	
	fst 4	−0.8993	0.7052	0.202		0.1446	0.6511	0.824		−0.3969	0.5691	0.486	
	Z2	0.4232	0.3779	0.263		0.3400	0.3576	0.342		0.4111	0.2743	0.134	
	Z3	0.3594	0.4061	0.376		−0.0076	0.3961	0.985		0.2010	0.2986	0.501	
C. coriaceus	interc.	−1.7804	0.5022	0.000	***	−2.3671	0.5774	0.000	***	−1.3491	0.3950	0.001	***
	fst 2	1.1268	0.6541	0.085	.	1.3251	0.7374	0.072	.	1.3132	0.5043	0.009	**
	fst 3	1.7444	0.7056	0.013	*	2.0437	0.6780	0.000	***	2.3143	0.5064	0.000	***
	fst 4	−18.3099	9565.85	0.998		−15.0008	3718.91	0.997		−18.0114	8434.89	0.998	
	Z2	−1.3491	0.4878	0.006	**	−1.8893	0.4555	0.000	***	−1.6282	0.3704	0.000	***
	Z3	−1.7433	0.5535	0.002	**	−2.5402	0.5014	0.000	***	−2.0328	0.4078	0.000	***
C. hortensis	interc.	−1.2211	0.4309	0.005	**	−1.5070	0.0513	0.006	**	−0.6934	0.4057	0.087	.
	fst 2	0.3203	0.6390	0.616		−0.0189	0.0858	0.982		0.2843	0.5953	0.633	
	fst 3	1.2173	0.6624	0.066	.	2.0460	0.7610	0.007	**	1.6898	0.5911	0.004	**
	fst 4	−16.5116	9282.76	0.999		−15.7900	10,690.00	0.999		−16.7419	9698.62	0.999	
	Z2	−0.8988	0.5207	0.084	.	−1.6910	0.5020	0.001	***	−1.2569	0.3776	0.001	***
	Z3	−3.8908	1.1116	0.000	***	−4.6770	1.0710	0.000	***	−4.2323	0.7833	0.000	***
C. violaceus	interc.	−0.3880	0.2243	0.083	.	−0.7902	0.2895	0.006	**	0.1213	0.1835	0.509	
	fst 2	−0.8097	0.5280	0.125		−0.8244	0.7313	0.260		−1.0273	0.4775	0.032	*
	fst 3	−0.4747	0.5967	0.426		−1.2281	0.8164	0.133		−0.9516	0.5377	0.077	.
	fst 4	−1.1461	0.7571	0.130		−1.3330	0.8914	0.135		−1.2776	0.6122	0.037	*
	Z2	−0.1240	0.5418	0.819		0.7072	0.7213	0.327		0.3956	0.4871	0.417	
	Z3	−0.1228	0.5913	0.835		1.2444	0.7907	0.116		0.6598	0.5298	0.213	

Finally, the model parameters were used to predict the species- and sex-specific activity densities of the Carabid species (except *Cal. inquisitor*). This procedure allowed predictions for combination of the different effects occurring at the level of forest stand type and the smaller-scale effects of the tree zones. The heatmaps created (Figure 3) revealed higher activity densities of C. *arvensis* predicted for the evenly mixed oak–pine stands (fst2) and the pine stands with oak proportions of <10% (fst3), including a low affinity to the oak effect zone. The model predictions for C. *coriaceus* and C. *hortensis* revealed clear activity density gradients. Higher densities were shown for the pine stands with <10% oak (fst3), declining towards the pure oak stand type (fst1). The higher activity densities of C. *coriaceus* and C. *hortensis* occurred in combination with a concentration of individuals within the oak tree effect zone (Z1). The activity density gradients were more pronounced for C. *coriaceus* and C. *hortensis* than for C. *arvensis*. In the case of C. *violaceus*, higher activity densities of imagines and females were predicted for the pure oak forest stand type in combination with the oak effect zone, with males possessing a higher affinity to the pine-related effect zones (Z3) across all forest stand types.

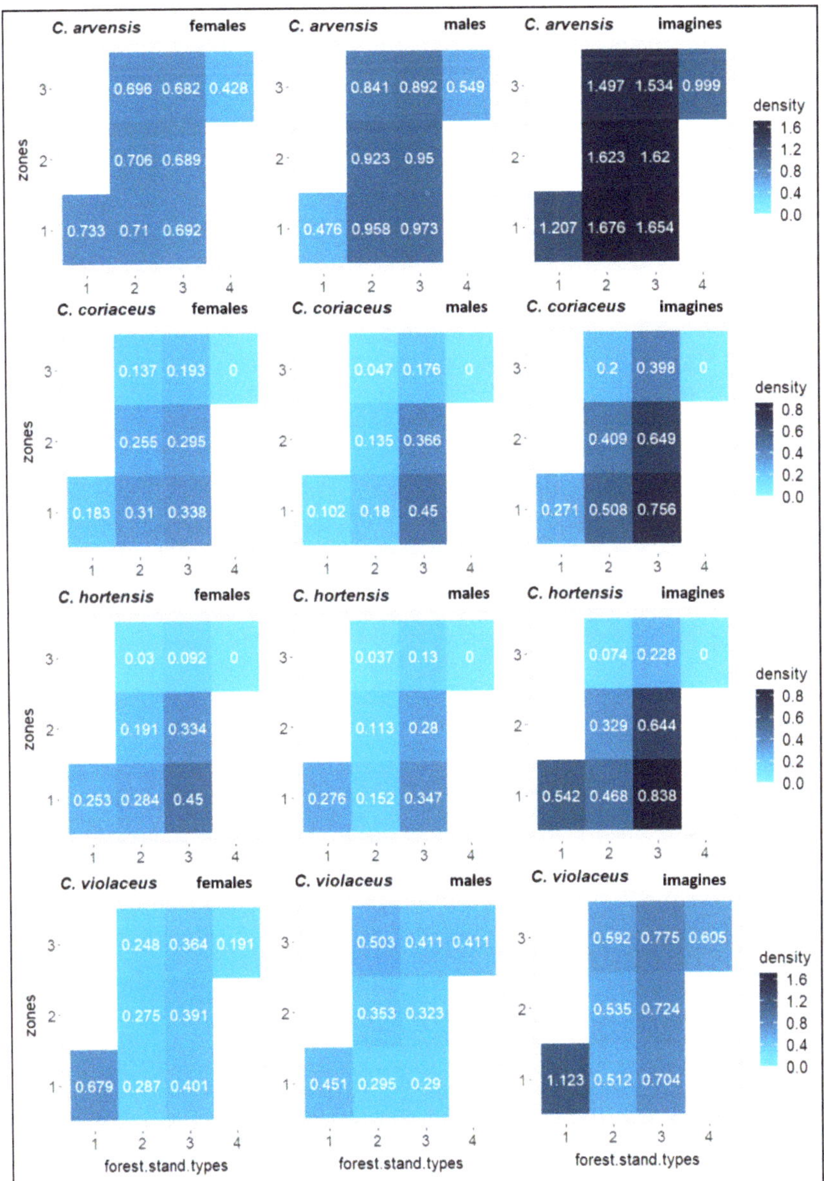

Figure 3. Carabus species- and sex-specific density (n/m^2) model predictions separated by forest stand type (fst1—pure oak, fst2—oak-pine mixture, fst3—<10% oak, fst4—pure pine) and tree species effect zone (Z1—oak, Z2—oak–pine mixture, Z3—pine). Note the differences between the density scales for the various Carabid species.

4. Discussion

4.1. Spatial Scale of Pure and Mixed Forest Stand Types

Regarding the aforementioned importance of carabids' role as indicators, as well as their model function, [18,25] referred to at the beginning of the paper, most of the relevant studies to date were concerned with structural diversity within the landscape [29,71].

Forests, as carabid habitats, have often been studied in comparison with open field conditions and forest patches [72,73]. The edge conditions lead to a highly selective categorisation of carabid species, given the considerable differences in habitat conditions between open fields and forest patches [74–78]. In forest ecosystems with less pronounced environmental gradients (e.g., transition zones) between structural elements, it is considerably more difficult to find clear differences in habitat conditions and in the resultant preferences of carabid species [79–81]. In order to develop multiple-use forest management with integrative conservation strategies, e.g., for carabids in central European forests, it is necessary to identify relevant spatial units or effect distances [36,40,82]. The definition of relevant structural elements (e.g., tree species compositions, dead wood, canopy gaps) as small-scale units within forests also applies to the characterisation of the habitat preferences of the different carabid species associated with forest ecosystems [22,28,83]. The distinction between pure and mixed forest stand types, as used in this study, is a first simple step, with a high degree of practical relevance, towards differentiating between the forest habitats of carabids [2,33,84,85]. It is worth noting here that traditional forest inventories classify forest stands with a proportion of <10% admixed tree species as pure stands [86]. This would be the case for the pine-dominated stand type with a low oak proportion (fst3) considered as part of this study and the ecological effects of the admixed old oak trees would be neglected as a result.

The assessment of forest types for carabids appears complex, given the importance of combining beetle parameters like activity density, body size and sex ratio, all of which provide indications of the habitat suitability and the vitality and the composition of the carabid species populations [87–89]. The species-specific comparison of carabids on the basis of forest stand type undertaken here revealed no significant effects on activity densities or on body sizes, although Skłodowski [22] observed increasing body sizes of carabids with an increasing proportion of deciduous tree species in forest stands. Previous studies also documented species-specific sex ratios as a function of different habitat conditions and seasonality [61,90]. The appreciably higher activity density (Figure 2) of the otherwise more immobile females (e.g., *C. violaceus* in pure oak) can be interpreted as a sign of suitable habitat conditions (oviposition site) [91]. Other studies, for example of *Pterostichus oblongopunctatus* (Fabricius, 1787), assumed a predominance of females to be indicative of better habitat quality [84]. However, other authors regarded a female-biased sex ratio to be a sign of unsuitable or disturbed habitats and increasing activity of females caused by hunger [47,61]. Focusing only on a single beetle parameter (density, body size or sex ratio) may lead to contrary assumptions for habitat derivation in carabids (e.g., *C. violaceus*).

The habitat preference of *Cal. inquisitor* was clear, being present exclusively in pure oak stands, where it occurred in high activity densities and with a high proportion of males. Although *Cal. inquisitor* is known as a species with a good capacity for flight, no individuals were recorded within the mixed oak–pine forest type with oak crown cover proportions of >30%. Similar results depicting the close affinity of *Cal. inquisitor* for oak trees were documented by du Bus de Warnaffe and Dufrêne [14]. In contrast to most ground beetles, *Cal. inquisitor* and *Cal. sycophanta* (L., 1758) have a strong affinity to deciduous trees as habitat [92]. The complete absence of *C. coriaceus* and *C. hortensis* in pure pine stands confirmed for both carabid species the basic assumption that forest stand types with an admixture of deciduous tree species such as sessile oak are preferable [21,84,93]. Moreover, the scarcity of carabid species within very homogenous, single-layered and pure pine stands was demonstrated by the fact that three (*Cal. inquisitor, C. coriaceus, C. hortensis*) of the five carabid species were absent (Figure 4). Contradictory findings presented by Barsoum et al. [33] indicated that the proportion of admixed oak trees in pine stands has no effect on the diversity of carabid species. Comparable activity densities of *C. coriaceus* and *C. hortensis* were recorded in pure oak stands and within mixed forest stand types. Only minor differences were revealed for the body sizes and the sex ratios of both species, with the exception of a higher female proportion of *C. hortensis* in the oak–pine forest stand type. No differences in the habitat preferences could be described for *C. coriaceus* and *C. hortensis* at the level of forest stand type without including smaller scale forest stand parameters such as tree effect zones (Figure 4). Interspecific competition between *C. coriaceus* and *C. hortensis* can be ruled out in spite of their similar forest type preferences, because of the relatively low activity densities [25,94]. *C. arvensis* and *C. violaceus* reached high activity densities within all four forest stand types. The use of such a broad range of forest habitats was already proven for *C. arvensis*, a species highly abundant in coniferous as well as deciduous forests, whereas *C. violaceus* is dominant in pure pine forests, subdominant in mixed coniferous forests and only an accessory species in oak–lime–hornbeam [84]. However, Magura et al. [95] found *C. violaceus* to be more pronounced in deciduous forests [95]. The differences in body sizes, related to forest stand type and sex-specific, was recorded for both species (females > males) [96]. This was also true for the sex ratios, which described increasing predominance of males with increasing proportions of pine. Females of both species were associated more with higher proportions of oak trees. As previously stated, the significance of the role of stand type in relation to sex ratios is debated [61,88]. No conclusive statements in relation to the habitat preferences of *C. arvensis* or *C. violaceus* were possible based on the forest stand type and the beetle parameters recorded.

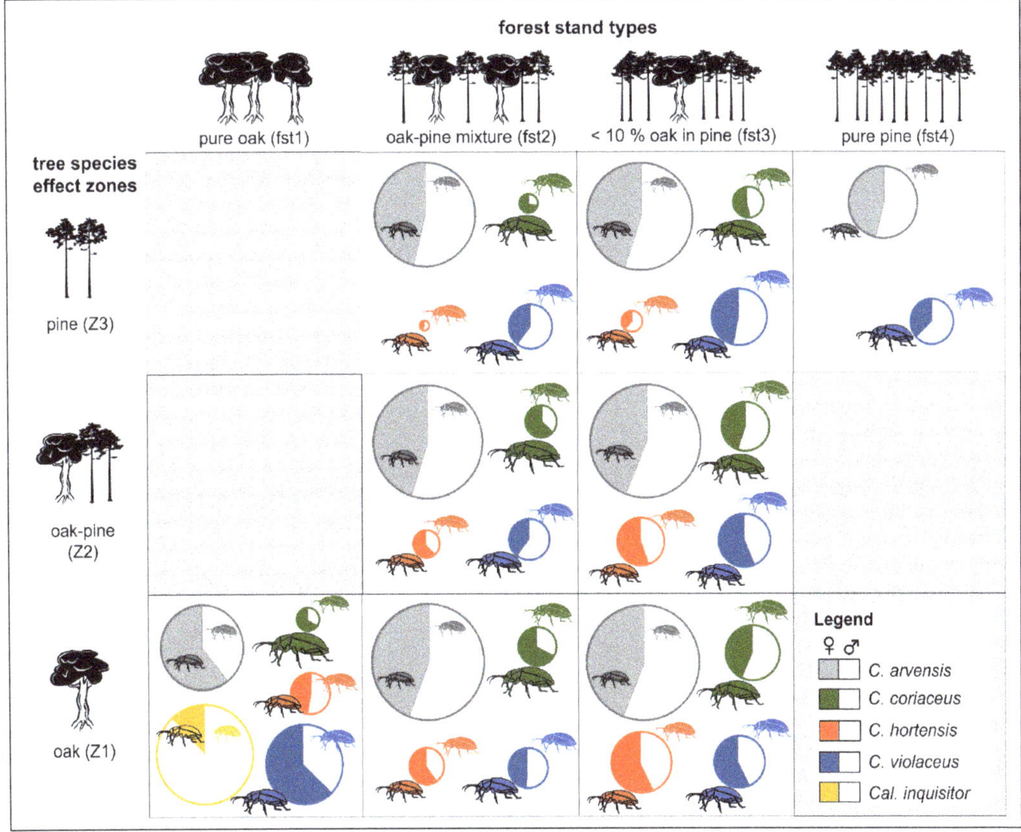

Figure 4. A schematic summary of the results depicting the effects of the combination of different scales (forest stand types and spatial effect zones) for *C. arvensis*, *C. coriaceus*, *C. hortensis*, *C. violaceus* and *Cal. inquisitor* separated by sex. The size of the circles represents the activity density of females and males. The relative body sizes of the beetles are described by the pictograms.

4.2. Spatial Scale of Single Trees and their Ecological Effect Zones

In order to improve the reliability of the information on the habitat preferences of carabid species in forests, additional data beyond the level of mere forest stand type (Figure 4) are necessary [14,27,33,79,97]. Structural heterogeneity within single-layered forest stands is mainly influenced by the proportions and spatial distributions of the different tree species present [30,32,56,98–100]. In this study, the pure oak and pine forest stand types were characterised by aggregation indices above 1, indicating a regular tree species distribution. Often small-scale structures of particular relevance for different carabid species, such as dead wood aggregations [101,102], habitat/retention trees [103,104] and canopy gaps [105,106] are absent from the pure stands which are managed primarily for wood production [83,107]. *Cal. inquisitor* was the only carabid species in a pure stand type to benefit from a more regular spatial distribution of oak trees, showing an increase in the activity density of individuals. The added value of adopting single tree-based effect zones is that it provided a more precise definition of potential niches for carabids, characterised by pure oak, mixed oak–pine and pure pine conditions within mixed forest stand types [33]. The results obtained for the mixed forest stand types revealed that the spatial heterogeneity of these zones is a decisive factor influencing the presence of carabids (Figure 4). As was shown in Table 2, the transition zone (proportion of oak–pine zone approx. 70%) was dominant within the oak–pine forest stand type (fst2), whereas Z2 attained a proportion of only 15% to 48% in the forest stand type with <10% oak (fst3). The proportion of niches was less balanced than is actually implied by the number of oak and pine trees within the mixed forest stand types, because the spatial manifestation of individual tree effect zones depends on tree traits [42,108,109]. In forest stand types with an equal number of oak and pine trees mixed amongst each other, the effect exerted by oak can be more intense for carabids, because of the greater impact of the high quantities of litter deposited seasonally on the soil surface [94,110]. Using spatial effect zones, more detailed niche differentiations are possible. *C. hortensis* responded more strongly to an increase in oak crown cover and oak effect zones. This manifested itself in higher individual densities and larger body sizes compared to *C. coriaceus*. The site preference of *C. hortensis*, identified at a microscale, may be explained by a greater affinity for more humid places [111] covered by mosses [112]. These microsites exist along the drip line between oak crown edges and pine trees [113]. The negative response to higher pine proportions and related effects is also stronger for *C. hortensis*. Effects of the spatial distribution of tree species and effect zones could be shown for both *C. hortensis* and *C. coriaceus* but in different ways. *C. hortensis* benefitted from a more regular distribution of oak trees, but this was less relevant for *C. coriaceus*.

No differences were observed between the sexes of *C. arvensis* in terms of forest stand type preference—in spite of the integration of smaller spatial units such as single tree effect zones. The reason for this was that this species is highly variable in its responses to different habitat conditions. This is reflected, for example, by the colour and metallic lustre of the beetle's body, a positive adaptation to differences in light availability [96]. The more shaded surroundings of oak trees situated within the oak–pine mixed stands had no obvious effect on the small-scale presence of *C. arvensis*. This may have been due to the relatively high light availability under the mixed oak–pine forest canopies and the slight zone-dependent light gradient between oak and pine trees [114,115]. To test for a potential affinity of *C. arvensis* to variable light conditions as affected by oak–pine mixtures in a forest stand, it would be useful to include the time of bud burst of oak trees as an additional seasonality aspect in future studies incorporating this spring-active carabid species [22,116].

Sex-specific responses to habitat conditions have been documented for various insect species [61,72]. In this study, *C. violaceus* showed the greatest sex-specific response in relation to single trees and effect zones, expressed in the activity density of individuals and in body size [25,51,117]. The females of *C. violaceus* revealed a high affinity for oak tree effects and avoided pine trees. The males, by contrast, were positively correlated with pine trees and the related effect zones, whereas oak trees were only preferred as spatial aggregates [98]. Only with an increase in the proportion of the oak–pine mixed zone (Z2) was a decrease observed in the activity density of individuals of both sexes of *C. violaceus*. Previous studies testing the effects of beech litter in spruce forests found only weak differences in the activity density of *C. violaceus* individuals [95], but the importance of sex-specific substrate preferences was also described for *Carabus hungaricus* (Fabricius, 1792) [52] and *Carabus clathratus* (L., 1761) [51]. The question of the temporal continuity of habitat uses in small-scale forest habitats, as described for the sexes of *C. violaceus*, can only be answered by including life cycle aspects for example, sex-specific behaviour during the reproductive period [47,91,118,119]. The importance of considering sex-specific smaller-scale habitat preferences to assess the suitability of forests is highlighted by the case of *C. violaceus*.

5. Conclusions

The derivation of the habitat preferences of carabid beetles in managed, pure, and mixed oak and pine forest stand types was successfully demonstrated in the study. The more unambiguous results of species- and sex-specific carabid parameters as responses to tree species effect zones illustrated the particular importance of small-scale spatial analyses, especially in mixed forests. Spatial information about tree species effect zones increases the accuracy of the identification of the habitat conditions in mixed forests preferred by carabid beetles. This approach represents an option to identify spatially optimal habitats [72] but also the limits to the extent that tree species combinations in mixed forests are suited to creating high variation of suitable habitat conditions for particular target species, for example, carabid beetles. Traditional silvicultural management activities, such as the targeted aggregation of oak trees in pine stands for the favourable impact this has on timber quality, can be evaluated on the basis of their impact on the different carabid species [98,120–122].

For carabid beetles to serve effectively as indicators or model species in spatially complex systems such as mixed forests, additional information about the continuity of environmental conditions within the corresponding microhabitats is required [15]. This applies in particular to species- and sex-specific analyses of carabid beetles. Additional research is also necessary to include the responses to tree species-related seasonality (deciduous versus evergreen coniferous trees) in mixed forests and life cycle aspects for carabid beetles [123]. Consideration should also be given to whether the more immobile and less chitinised larvae of carabids might be more suitable as environmental indicators [17,18,91,124].

While the results presented here are valid for pure and mixed forests consisting of oak and pine, our findings cannot be directly applied to other tree species constellations, because the spatial effect zones are strictly linked to the tree species and individual tree traits. Further research focusing explicitly on spatio-temporal interactions between tree species constellations in mixed forests and the effect on carabid species would be beneficial.

Author Contributions: Data collecting and data curation, A.W.; conceptualization, A.W. and F.H.; methodology; software; validation; formal analysis; writing—original draft preparation; writing—review and editing, A.W., S.W. and F.H.; visualisation, A.W. and F.H.; supervision, S.W. All authors have read and agreed to the published version of the manuscript.

Funding: This research was funded by a private scholarship provided by the Michael-Jahr-Foundation. Open Access Funding by the Publication Fund of the TU Dresden.

Institutional Review Board Statement: Not applicable.

Data Availability Statement: The data presented in this study are available on request from the corresponding author, because the data are part of a Ph.D. Thesis (A.W.) not yet finished.

Acknowledgments: The results presented are part of a wider study into single-tree research. The authors would like to express especial thanks to the Michael-Jahr-Foundation for supporting the whole study with a scholarship. We would also like to thank the companies Gebr. Ostendorf

Kunst-stoffe GmbH and Co. KG and Lentia Pirna GmbH for supporting our study by supplying materials for the pitfall traps. We are grateful to Landesbetrieb Forst Brandenburg (LFB) and Landeskompetenzzentrum Forst Eberswalde (LFE) for providing climate data. A special thanks goes to Antje Karge, Michael Wehnert-Kohlenbrenner, Andreas Möhring and Lumir Dobrovolný for helping during stem position measurements and to our proofreader David Butler Manning.

Conflicts of Interest: The authors declare no conflict of interest.

References

1. Brockerhoff, E.G.; Jactel, H.; Parrotta, J.A.; Quine, C.P.; Sayer, J. Plantation forests and biodiversity: Oxymoron or opportunity? *Biodivers. Conserv.* **2008**, *17*, 925–951. [CrossRef]
2. Oxbrough, A.; French, V.; Irwin, S.; Kelly, T.C.; Smiddy, P.; O'Halloran, J. Can mixed species stands enhance arthropod diversity in plantation forests? *For. Ecol. Manag.* **2012**, *270*, 11–18. [CrossRef]
3. Bravo-Oviedo, A.; Pretzsch, H.; del Río, M. *Dynamics, Silviculture and Management of Mixed Forests*; Managing Forest Ecosystems Vol 31; Springer: Cham, Switzerland, 2018.
4. Anders, S.; Beck, W.; Bolte, A.; Hofmann, G.; Jenssen, M.; Krakau, U.; Müller, J. *Ökologie und Vegetation der Wälder Nordostdeutschlands.—Einfluß von Niederschlagsarmut und erhöhtem Stickstoffeintrag auf Kiefern-, Eichen- und Buchen-Wald- und Forstökosysteme des Nordostdeutschen Tieflandes*, 1st ed.; Verlag Dr. Kessel: Oberwinter, Germany, 2002.
5. Goris, R.; Kint, V.; Haneca, K.; Geudens, G.; Beeckman, H.; Verheyen, K. Long-term dynamics in a planted conifer forest with spontaneous ingrowth of broad-leaved trees. *Appl. Veg. Sci.* **2007**, *10*, 219–228. [CrossRef]
6. Pretzsch, H.; Steckel, M.; Heym, M.; Biber, P.; Ammer, C.; Ehbrecht, M.; Bielak, K.; Bravo, F.; Ordóñez, C.; Collet, C.; et al. Stand growth and structure of mixed-species and monospecific stands of Scots pine (*Pinus sylvestris* L.) and oak (*Q. robur* L., *Quercus petraea* (Matt.) Liebl.) analysed along a productivity gradient through Europe. *Eur. J. For. Res.* **2020**, *139*, 349–367. [CrossRef]
7. Bravo-Oviedo, A.; Pretzsch, H.; Ammer, C.; Andenmatten, E.; Barbati, A.; Barreiro, S.; Brang, P.; Bravo, F.; Coll, L.; Corona, P.; et al. European Mixed Forests: Definition and research perspectives. *For. Syst.* **2014**, *23*, 518–533. [CrossRef]
8. Jactel, H.; Brockerhoff, E.G. Tree diversity reduces herbivory by forest insects. *Ecol. Lett.* **2007**, *10*, 835–848. [CrossRef] [PubMed]
9. Knoke, T.; Ammer, C.; Stimm, B.; Mosandl, R. Admixing broadleaved to coniferous tree species: A review on yield, ecological stability and economics. *Eur. J. For. Res.* **2008**, *127*, 89–101. [CrossRef]
10. Szmyt, J.; Tarasiuk, S. Species-specific spatial structure, species coexistence and mortality pattern in natural, uneven-aged Scots pine (*Pinus sylvestris* L.)-dominated forest. *Eur. J. For. Res.* **2018**, *137*, 1–16. [CrossRef]
11. Butterfield, J.; Benitez Malvido, J. Effect of mixed-species tree planting on the distribution of soil invertebrates. In *The Ecology of Mixed-Species Stands of Trees*; Cannell, M.G.R., Malcolm, D.C., Robertson, P.A., Eds.; Special Publication Number 11 of the British Ecological Society; Oxford-Blackwell Scientific Publications: London, UK, 1992; pp. 255–265.
12. Gamfeldt, L.; Snäll, T.; Bagchi, R.; Jonsson, M.; Gustafsson, L.; Kjellander, P.; Ruiz-Jaen, M.C.; Fröberg, M.; Stendahl, J.; Philipson, C.D.; et al. Higher levels of multiple ecosystem services are found in forests with more tree species. *Nat. Commun.* **2013**, *4*, 1340. [CrossRef]
13. Sobek, S.; Steffan-Dewenter, I.; Scherber, C.; Tscharntke, T. Spatiotemporal changes of beetle communities across a tree diversity gradient. *Divers. Distrib.* **2009**, *15*, 660–670. [CrossRef]
14. du Bus de Warnaffe, G.; Dufrêne, M. To what extent can management variables explain species assemblages? A study of carabid beetles in forests. *Ecography* **2004**, *27*, 701–714. [CrossRef]
15. Lassau, S.A.; Hochuli, D.F.; Cassis, G.; Reid, C.A.M. Effects of Habitat Complexity on Forest Beetle Diversity: Do Functional Groups Respond Consistently? *Divers. Distrib.* **2005**, *11*, 73–82. [CrossRef]
16. Rainio, J.; Niemelä, J. Ground beetles (Coleoptera: Carabidae) as bioindicators. *Biodivers. Conserv.* **2003**, *12*, 487–506. [CrossRef]
17. Work, T.T.; Koivula, M.; Klimaszewski, J.; Langor, D.; Sweeney, J.; Hébert, C. Evaluation of carabid beetles as indicators of forest change in Canada. *Can. Entomol.* **2008**, *140*, 393–414. [CrossRef]
18. Koivula, M. Useful model organisms, indicators, or both? Ground beetles (Coleoptera, Carabidae) reflecting environmental conditions. *ZooKeys* **2011**, *100*, 287–317. [CrossRef]
19. Walsh, P.J.; Day, K.R.; Leather, S.R.; Smith, A. The Influence of Soil Type and Pine Species on the Carabid Community of a Plantation Forest with a History of Pine Beauty Moth Infestation. *Forestry* **1993**, *66*, 135–146. [CrossRef]
20. Pearce, J.L.; Venier, L.A. The use of ground beetles (Coleoptera: Carabidae) and spiders (Araneae) as bioindicators of sustainable forest management: A review. *Ecol. Indic.* **2006**, *6*, 780–793. [CrossRef]
21. Lindroth, C.H. *Ground Beetles (Carabidae) of Fennoscandia. A Zoogeographic Study. Part 1: Specific Knowledge Regarding the Species. Translation of: Die fennoskandischen Carabidae: Eine Tiergeographische Studie I. Spezieller Teil*; Amerind Publishing Co. Pvt. Ltd.: New Delhi, India, 1992.
22. Skłodowski, J.J.W. Interspecific body size differentiation in Carabus assemblages in the Białowieża Primeval Forest, Poland. In Proceedings of the 11th European Carabidologists' Meeting, Århus, Denmark, 21–24 July 2003; Lövei, G.L., Toft, S., Eds.; DIAS Report, No. 114. Ministry of Food, Agriculture and Fisheries: Tjele, Denmark, 2005; pp. 291–303.
23. Dahl, F. *Die Tierwelt Deutschlands und der Angrenzenden Meeresteile nach ihren Merkmalen und Lebensweise 7.Teil: Coleoptera oder Käfer I: Carabidae (Laufkäfer)*, 1st ed.; Verlag von Gustav Fischer: Jena, Germany, 1928.

24. Turin, H.; Penev, L.; Casale, A. *The Genus Carabus in Europe. A Synthesis*; Fauna Europaea Evertebrata No. 2; Pensoft Publishers & European Invertebrates Survey: Sofia, Bulgaria, 2003.
25. Thiele, H.-U. *Carabid Beetles in Their Environments: A Study on Habitat Selection by Adaptations in Physiology and Behaviour*; Zoophysiology and Ecology Volume 10; Springer: Berlin/Heidelberg, Germany, 1977.
26. Stork, N.E. *The Role of Ground Beetles in Ecological and Environmental Studies*; Intercept Ltd.: Andover, MA, USA, 1990.
27. Sroka, K.; Finch, O.-D. Ground beetle diversity in ancient woodland remnants in north-western Germany (Coleoptera, Carabidae). *J. Insect Conserv.* **2006**, *10*, 335–350. [CrossRef]
28. Lange, M.; Türke, M.; Pašalić, E.; Boch, S.; Hessenmöller, D.; Müller, J.; Prati, D.; Socher, S.A.; Fischer, M.; Weisser, W.W.; et al. Effects of forest management on ground-dwelling beetles (Coleoptera; *Carabidae, Staphylinidae*) in Central Europe are mainly mediated by changes in forest structure. *For. Ecol. Manag.* **2014**, *329*, 166–176. [CrossRef]
29. Taboada, A.; Kotze, D.J.; Tárrega, R.; Salgado, J.M. Traditional forest management: Do carabid beetles respond to human-created vegetation structures in an oak mosaic landscape? *For. Ecol. Manag.* **2006**, *237*, 436–449. [CrossRef]
30. Clarke, R.T.; Thomas, J.A.; Elmes, G.W.; Hochberg, M.E. The Effects of Spatial Patterns in Habitat Quality on Community Dynamics within a Site. *Proc. R. Soc. Lond. B* **1997**, *264*, 347–354. [CrossRef]
31. Tomppo, E.; Gschwantner, T.; Lawrence, M.; McRoberts, R.E. *National Forest Inventories. Pathways for Common Reporting*; Springer Science + Business Media B.V.: New York, NY, USA, 2010.
32. Niemelä, J. Invertebrates and Boreal Forest Management. *Conserv. Biol.* **1997**, *11*, 601–610. [CrossRef]
33. Barsoum, N.; Fuller, L.; Ashwood, F.; Reed, K.; Bonnet-Lebrun, A.-S.; Leung, F. Ground-dwelling spider (Araneae) and carabid beetle (Coleoptera: Carabidae) community assemblages in mixed and monoculture stands of oak (*Quercus robur L./Quercus petraea* (Matt.) Liebl.) and Scots pine (*Pinus sylvestris* L.). *For. Ecol. Manag.* **2014**, *321*, 29–41. [CrossRef]
34. Wu, H.; Sharpe, P.J.H.; Walker, J.; Penridge, L.K. Ecological Field Theory: A Spatial Analysis of Resource Interference among Plants. *Ecol. Model.* **1985**, *29*, 215–243. [CrossRef]
35. Pretzsch, H. Canopy space filling and tree crown morphology in mixed-species stands compared with monocultures. *For. Ecol. Manag.* **2014**, *327*, 251–264. [CrossRef]
36. Laca, E.A. Multi-Scape Interventions to Match Spatial Scales of Demand and Supply of Ecosystem Services. *Front. Sustain. Food Syst.* **2021**, *4*, 607276. [CrossRef]
37. Kuuluvainen, T.; Pukkala, T. Effect of Scots pine seed trees on the density of ground vegetation and tree seedlings. *Silva Fenn.* **1989**, *23*, 159–167. [CrossRef]
38. Schua, K.; Fischer, H.; Lehmann, B.; Wagner, S. Single tree effects of sessile oak (*Quercus petraea* (Matt.) Liebl.) within pure stands (*Pinus sylvestris* L.) on topsoil properties. *Allg. For. J. Ztg.* **2007**, *178*, 172–179, (In German with English abstract).
39. Gruba, P.; Mulder, J.; Pacanowskia, P. Combined effects of soil disturbances and tree positions on spatial variability of soil pH_{CaCl2} under oak and pine stands. *Geoderma* **2020**, *376*, 114537. [CrossRef]
40. Wagner, S.; Herrmann, I.; Huth, F. Tools familiar, impact unexpected: Silviculture and ecosystem services on a small forest scale. *Allg. For. J. Ztg.* **2020**, *190*, 89–100. [CrossRef]
41. Jonard, M.; Andre, F.; Ponette, Q. Tree species mediated effects on leaf litter dynamics in pure and mixed stands of oak and beech. *Can. J. For. Res.* **2008**, *38*, 528–538. [CrossRef]
42. Wehnert, A.; Wagner, S. Niche partitioning in carabids: Single-tree admixtures matter. *Insect Conserv. Divers.* **2019**, *12*, 131–146. [CrossRef]
43. Beniamino, F.; Ponge, J.F.; Arpin, P. Soil acidification under the crown of oak trees. I. Spatial distribution. *For. Ecol. Manag.* **1991**, *40*, 221–232. [CrossRef]
44. Abella., S.R.; Springer, J.D. Canopy-tree influences along a soil parent material gradient in *Pinus ponderosa-Quercus gambelii* forests, northern Arizona. *J. Torrey Bot. Soc.* **2008**, *135*, 26–36. [CrossRef]
45. Hassan, S.M.T.; Ghimire, C.P.; Lubczynski, M.W. Remote sensing upscaling of interception loss from isolated oaks: Sardon catchment case study, Spain. *J. Hydrol.* **2017**, *555*, 489–505. [CrossRef]
46. Kádár, K.; Fazekas, J.P.; Sárospataki, M.; Lövei, G.L. Seasonal dynamics, age structure and reproduction of four Carabus species (Coleoptera: Carabidae) living in forested landscapes in Hungary. *Acta Zool. Hung.* **2015**, *61*, 57–72. [CrossRef]
47. Paoletti, M.G.; Martin Cantarino, C. Sex ratio alterations in terrestrial woodlice populations (Isopoda: Oniscidea) from agroecosystems subjected to different agricultural practices in Italy. *Appl. Soil Ecol.* **2002**, *19*, 113–120. [CrossRef]
48. Knapp, M.; Knappová, J. Measurement of body condition in a common carabid beetle, *Poecilus cupreus*: A comparison of fresh weight, dry weight, and fat content. *J. Insect Sci.* **2013**, *13*, 6. [CrossRef] [PubMed]
49. Jervis, M.; Kidd, N. *Insect Natural Enemies. Practical Approaches to Their Study and Evaluation*; Chapman & Hall: London, UK, 1996.
50. Leather, S. *Insect Sampling in Forest Ecosystems*; Blackwell Science Ltd: Malden, MA, USA, 2005.
51. Huk, T.; Kühne, B. Substrate selection by *Carabus clatratus* (Coleoptera, Carabidae) and its consequences for offspring development. *Oecologia* **1999**, *121*, 348–354. [CrossRef] [PubMed]
52. Pokluda, P.; Hauck, D.; Cizek, L. Importance of marginal habitats for grassland diversity: Fallows and overgrown tall-grass steppe as key habitats of endangered ground-beetle *Carabus hungaricus*. *Insect Conserv. Divers.* **2012**, *5*, 27–36. [CrossRef]
53. LFB and LFE. 2017. Available online: http://www.forstliche-umweltkontrolle-bb.de/r3_meteo.php (accessed on 6 November 2017).

54. Müller, J.K. Die Bedeutung der Fallenfang-Methode für die Lösung ökologischer Fragestellungen. *Zool. Jb. Syst.* **1984**, *111*, 281–305.
55. Brown, G.R.; Matthews, I.M. A review of extensive variation in the design of pitfall traps and a proposal for a standard pitfall trap design for monitoring ground-active arthropod biodiversity. *Ecol. Evol.* **2016**, *6*, 3953–3964. [CrossRef] [PubMed]
56. Ziesche, M. Effect of Forest Structure and Small-Scale Environmental Conditions on the Community of Epigeic Arthropods (Carabidae, Araneae). Ph.D. Thesis, Technische Universität Dresden, Fakultät Umweltwissenschaften, Tharandt, Germany, 2016.
57. Heydemann, B. Agrarökologische Problematik (dargetan an Untersuchungen über die Tierwelt der Bodenoberfläche der Kulturfelder). Ph.D. Thesis, Hohen Philosophischen Fakultät an der Christian-Albrechts-Universität, Kiel, Germany, 1953.
58. Scheller, H.V. Pitfall trapping as the basis for studying ground beetle (Carabidae) predation in spring barley. *Tidsskr. Planteavl.* **1984**, *88*, 317–324.
59. Mühlenberg, M. *Freilandökologie*, 3rd ed.; Quelle & Meyer Verlag: Heidelberg/Wiesbaden, Germany, 1993.
60. Langraf, V.; David, S.; Babosová, R.; Petrovičová, K.; Schlarmannová, J. Change of Ellipsoid Biovolume (EV) of Ground Beetles (Coleoptera, Carabidae) along an Urban–Suburban–Rural Gradient of Central Slovakia. *Diversity* **2020**, *12*, 475. [CrossRef]
61. Szyszko, J.; Gryuntal, S.; Schwerk, A. Differences in Locomotory Activity Between Male and Female *Carabus hortensis* (Coleoptera: Carabidae) in a Pine Forest and a Beech Forest in Relation to Feeding State. *Environ. Entomol.* **2004**, *33*, 1442–1446. [CrossRef]
62. Clark, P.J.; Evans, F.C. Distance to nearest neighbor as a measure of spatial relationships in populations. *Ecology* **1954**, *35*, 445–453. [CrossRef]
63. Krebs, C.J. *Ecological Methodology*, 2nd ed.; Addison Wesley Longman Inc., University of British Columbia: Vancouver, BC, Canada, 1999.
64. Baddeley, A.; Rubak, E.; Turner, R. *Spatial Point Patterns. Methodology and Applications with R*; Chapman & Hall/CRC Interdisciplinary Statistics Series; CRC Press: Boca Raton, FL, USA, 2016.
65. Schober, P.; Boer, C.; Schwarte, L.A. Correlation Coefficients: Appropriate Use and Interpretation. *Anesth. Analg.* **2018**, *126*, 1763–1768. [CrossRef]
66. Zinke, P.J. The Pattern of Influence of Individual Forest Trees on Soil Properties. *Ecology* **1962**, *43*, 130–133. [CrossRef]
67. Shamsutdinov, Z.; Ubaydullaev, S.; Shamsutdinov, N.; Mirzaev, B.; Mamatov, F.; Chorshabiyev, N. The concept of the phytogenic field: Theory, research experience and practical significance. *IOP Conf. Ser. Earth Env. For. Sci.* **2020**, *614*, 012164. [CrossRef]
68. Bolker, B.M.; Brooks, M.E.; Clark, C.J.; Geange, S.W.; Poulsen, J.R.; Stevens, M.H.H.; White, J.-S.S. Generalized linear mixed models: A practical guide for ecology and evolution. *Trends Ecol. Evol.* **2009**, *24*, 127–135. [CrossRef]
69. Vinatier, F.; Tixier, P.; Duyck, P.-F.; Lescourret, F. Factors and mechanisms explaining spatial heterogeneity: A review of methods for insect populations. *Methods Ecol. Evol.* **2011**, *2*, 11–22. [CrossRef]
70. Magnusson, A.; Skaug, H.; Nielsen, A.; Berg, C.; Kristensen, K.; Maechler, M.; van Bentham, K.; Bolker, B.; Sadat, N.; Lüdecke, D.; et al. glm TMB: Generalized Linear Mixed Models Using Template Model Builder. Date/Publication 2020-07-02 11:30:17 UTC. 2020. Available online: https://cran.r-project.org/web/packages/glmmTMB/glmmTMB.pdf (accessed on 19 January 2021).
71. Barbaro, L.; Pontcharraud, L.; Vetillard, F.; Guyon, D.; Jactel, H. Comparative responses of bird, carabid, and spider assemblages to stand and landscape diversity in maritime pine plantation forest. *Écoscience* **2005**, *12*, 110–121. [CrossRef]
72. Samways, M.J.; McGeoch, M.A.; New, T.R. *Insect Conservation: A Handbook of Approaches and Methods*; Techniques in Ecology & Conservation Series; Oxford University Press: London, UK, 2013.
73. Davis, D.E.; Gagné, S.A. Boundaries in ground beetle (Coleoptera: Carabidae) and environmental variables at the edges of forest patches with residential developments. *PeerJ* **2018**, *6*, e4226. [CrossRef] [PubMed]
74. Murcia, C. Edge effects in fragmented forests: Implications for conservation. *Tree* **1995**, *10*, 58–62. [CrossRef]
75. Didham, R.K. An overview of invertebrate responses to forest fragmentation. In *Forests and Insects*; Watt, A.D., Stork, N.E., Hunter, M.D., Eds.; Chapman & Hall: London, UK, 1997; pp. 305–320.
76. Hamberg, L.; Lehvävirta, S.; Kotze, D.J. Forest edge structure as a shaping factor of understorey vegetation in urban forests in Finland. *For. Ecol. Manag.* **2009**, *257*, 712–722. [CrossRef]
77. Ries, L.; Sisk, T.D. What is an edge species? The implications of sensitivity to habitat edges. *Oikos* **2010**, *119*, 1636–1642. [CrossRef]
78. Fahrig, L. Rethinking patch size and isolation effects: The habitat amount hypothesis. *J. Biogeogr.* **2013**, *40*, 1649–1663. [CrossRef]
79. Niemelä, J.; Haila, Y.; Halme, E.; Pajunen, T.; Punttila, P. Small-scale heterogeneity in the spatial distribution of carabid beetles in the southern Finnish taiga. *J. Biogeogr.* **1992**, *19*, 173–181. [CrossRef]
80. Day, K.R.; Marshall, S.; Heaney, C. Association between Forest Type and Invertebrates: Ground Beetle Community Patterns in a Natural Oakwood and Juxtaposed Conifer Plantation. *Forestry* **1993**, *66*, 37–50. [CrossRef]
81. Hunter, M.L. *Maintaining Biodiversity in Forest Ecosystems*; Cambridge University Press: Cambridge, UK, 1999.
82. Kraus, D.; Krumm, F. (Eds.) *Integrative Approaches as an Opportunity for the Conservation of Forest Biodiversity*; European Forest Institute: Freiburg, Germany, 2013.
83. Niemelä, J.; Koivula, M.; Kotze, D.J. The effects of forestry on carabid beetles (Coleoptera: Carabidae) in boreal forest. *J. Insect Conserv.* **2007**, *11*, 5–18. [CrossRef]
84. Szujecki, A. *Ecology of Forest Insects*; Series Entomologica; Dr. W. Junk Publishers: Dordrecht, The Netherlands, 1987; Volume 26.
85. Jukes, M.R.; Peace, A.J.; Ferris, R. Carabid beetle communities associated with coniferous plantations in Britain: The influence of site, ground vegetation and stand structure. *For. Ecol. Manag.* **2001**, *148*, 271–286. [CrossRef]

86. Pretzsch, H.; Forrester, D.I.; Bauhus, J. *Mixed-Species Forests: Ecology and Management*; Springer: Berlin/Heidelberg, Germany, 2017. [CrossRef]
87. Schowalter, T.D. *Insect Ecology. An Ecosystem Approach*, 2nd ed.; Elsevier: Amsterdam, The Netherlands, 2006.
88. Moya-Laraño, J.; Macías-Ordóñez, R.; Blanckenhorn, W.U.; Fernández-Montraveta, C. Analysing Body Condition: Mass, Volume or Density? *J. Anim. Ecol.* **2008**, *77*, 1099–1108. [CrossRef] [PubMed]
89. Langraf, V.; Petrovičová, K.; David, S.; Ábelová, M.; Schlarmannová, J. Body volume in ground beetles (Carabidae) reflects biotope disturbance. *Folia Oecologica* **2017**, *44*, 114–120. [CrossRef]
90. Loreau, M. Annual activity and life cycles of carabid beetles in two forest communities. *Holarct. Ecol.* **1985**, *8*, 228–235. [CrossRef]
91. Lövei, G.L.; Sunderland, K.D. Ecology and behavior of Ground Beetles (Coleoptera: Carabidae). *Annu. Rev. Entomol.* **1996**, *41*, 231–256. [CrossRef]
92. Müller-Motzfeld, G. Laufkäfer in Wäldern Deutschlands. *Angew. Carab. Suppl.* **2001**, *II*, 9–20.
93. Wehnert, A.; Wagner, S.; Huth, F. Spatio-Temporal Distribution of Carabids Influenced by Small-Scale Admixture of Oak Trees in Pine Stands. *Diversity* **2020**, *12*, 398. [CrossRef]
94. Shibuya, S.; Kubota, K.; Ohsawa, M.; Kikvidze, Z. Assembly rules for ground beetle communities: What determines community structure, environmental factors or competition? *Eur. J. Entomol.* **2011**, *108*, 453–459. [CrossRef]
95. Magura, T.; Tóthmérész, B.; Elek, Z. Effects of leaf-litter addition on carabid beetles in a non-native Norway spruce plantation. *Acta Zool. Acad. Sci. Hung.* **2004**, *50*, 9–23.
96. Tyler, G. Variability in colour, metallic lustre, and body size of *Carabus arvensis* Herbst, 1784 (Coleoptera: Carabidae) in relation to habitat properties. *Entomol. Fenn.* **2010**, *21*, 90–96. [CrossRef]
97. Jelaska, L.Š.; Dumbović, V.; Kučinić, M. Carabid beetle diversity and mean individual biomass in beech forests of various age. *ZooKeys* **2011**, *100*, 393–405. [CrossRef]
98. Ohsawa, M. The role of isolated old oak trees in maintaining beetle diversity within larch plantations in the central mountainous region of Japan. *For. Ecol. Manag.* **2007**, *250*, 215–226. [CrossRef]
99. Wiezik, M.; Svitok, M.; Dovčiak, M. Conifer introductions decrease richness and alter composition of litter-dwelling beetles (Coleoptera) in Carpathian oak forests. *For. Ecol. Manag.* **2007**, *247*, 61–71. [CrossRef]
100. Worthen, W.B.; Merriman, D.C.G. Relationships between Carabid Beetle Commmunities and Forest Stand Parameters: Taxon Congruencer or Habitat Association. *Southeast. Nat.* **2013**, *12*, 379–386. [CrossRef]
101. Skłodowski, J.; Szczeszek, J. Dead wood modifies mobility of ground beetles. *Baltic J. Coleopterol.* **2015**, *15*, 91–98.
102. Andrési, D.; Bali, L.; Tuba, K.; Szinetár, C. Comparative study of ground beetle and ground-dwelling spider assemblages of artificial gap openings. *Community Ecol.* **2018**, *19*, 133–140. [CrossRef]
103. Matveinen-Huju, K.; Niemelä, J.; Rita, H.; O´Hara, R.B. Retention-tree groups in clear-cuts: Do they constitute "life-boats" for spiders and carabids? *For. Ecol. Manag.* **2006**, *230*, 119–135. [CrossRef]
104. Koch Widerberg, M. Oak as Retention Tree in Commercial Spruce Forests. Effects on Species Diversity of Saproxylic Beetles and Wood Production. Ph.D. Thesis, Swedish University of Agricultural Sciences, Alnarp, Sweden, 2013. No. 2013:66.
105. Koivula, M.; Niemelä, J. Gap felling as a forest harvesting method in boreal forests: Responses of carabid beetles (Coleoptera, Carabidae). *Ecography* **2003**, *26*, 179–187. [CrossRef]
106. Ulyshen, M.D.; Hanula, J.L.; Horn, S.; Kilgo, J.C.; Moorman, C.E. The response of ground beetles (Coleoptera: Carabidae) to selection cutting in a South Carolina bottomland hardwood forest. *Biodivers. Conserv.* **2006**, *15*, 261–274. [CrossRef]
107. Vítková, L.; Bače, R.; Kjučukov, P.; Svoboda, M. Deadwood management in Central European forests: Key considerations for practical implementation. *For. Ecol. Manag.* **2018**, *429*, 394–405. [CrossRef]
108. Weber, P.; Bardgett, R.D. Influence of single trees on spatial and temporal patterns of belowground properties in native pine forest. *Soil Biol. Biochem.* **2011**, *43*, 1372e1378. [CrossRef]
109. Vrška, T.; Janík, D.; Pálková, M.; Adam, D.; Trochta, J. Below- and above-ground biomass, structure and patterns in ancient lowland coppices. *iForest* **2016**, *10*, 23–31. [CrossRef]
110. García, L.V.; Maltez-Mouro, S.; Pérez-Ramos, I.M.; Freitas, H.; Marañón, T. Counteracting gradients of light and soil nutrients in the understorey of Mediterranean oak forests. *Web Ecol.* **2006**, *6*, 67–74. [CrossRef]
111. Skłodowski, J. Consequence of the transformation of a primeval forest into a managed forest for carabid beetles (Coleoptera: Carabidae)—A case study from Białowieża (Poland). *Eur. J. Entomol.* **2014**, *111*, 639–648. [CrossRef]
112. Heliölä, J.; Koivula, M.; Niemelä, J. Distribution of Carabid Beetles (Coleoptera, Carabidae) across a Boreal Forest-Clearcut Ecotone. *Conserv. Biol.* **2001**, *15*, 370–377. [CrossRef]
113. Wagner, S.; Fischer, H.; Huth, F. Canopy effects on vegetation caused by harvesting and regeneration treatments. *Eur. J. For. Res.* **2011**, *130*, 17–40. [CrossRef]
114. Lintunen, A.; Kaitaniemi, P.; Perttunen, J.; Sievänen, R. Analysing species-specific light transmission and related crown characteristics of *Pinus sylvestris* and *Betula pendula* using a shoot-level 3D model. *Can. J. For. Res.* **2013**, *43*. [CrossRef]
115. Perot, T.; Mårell, A.; Korboulewsky, N.; Seigner, V.; Balandier, P. Modeling and predicting solar radiation transmittance in mixed forests at a within-stand scale from tree species basal area. *For. Ecol. Manag.* **2017**, *390*, 127–136. [CrossRef]
116. Grüm, L. Spatial Distribution of Males and Females of *Carabus arcensis* Hbst. in the Breeding Season. In *The Role of Ground Beetles in Ecological and Environmental Studies*; Stork, N.E., Ed.; Intercept Ltd.: Andover, MA, USA, 1990; pp. 277–287.
117. Loreau, M. Vertical distribution of activity of carabid beetles in a beech forest floor. *Pedobiologia* **1987**, *30*, 173–178.

118. Hawes, C.; Stewart, A.J.A.; Evans, H.F. The impact of wood ants (*Formica rufa*) on the distribution and abundance of ground beetles (Coleoptera: Carabidae) in a Scots pine plantation. *Oecologia* **2002**, *131*, 612–619. [CrossRef]
119. Luff, M.L. Biology and ecology of immature stages of ground beetles (Carabidae). In Proceedings of the 11th European Carabidologists' Meeting, Århus, Denmark, 21–24 July 2003; Lövei, G.L., Toft, S., Eds.; DIAS Report, No. 114. Ministry of Food, Agriculture and Fisheries: Tjele, Denmark, 2005; pp. 183–208.
120. Koivula, M.; Niemelä, J. Boreal Carabid Beetles (Coleoptera, Carabidae) in Managed Spruce Forests—A Summary of Finnish Case Studies. *Silva Fenn.* **2002**, *36*, 423–436. [CrossRef]
121. Krissl, K.; Müller, F. Zweckmäßige Dauermischungsformen und Mischungsregulierung. *Österr. Forstztg.* **1990**, *3*, 29–32.
122. Perot, T.; Picard, N. Mixture enhances productivity in a two-species forest: Evidence from a modeling approach. *Ecol. Res. Ecol. Soc. JPN.* **2012**, *27*, 83–94. [CrossRef]
123. Šiška, B.; Eliašová, M.; Kollár, J. Carabus Population Response to Drought in Lowland Oak Hornbeam Forest. *Water* **2020**, *12*, 3284. [CrossRef]
124. Arndt, E.; Arndt, M. Auswertung der Bodenfallenfänge von Carabidenlarven (Coleoptera) im Hakel (Nordharzvorland). *Hercynia* **1987**, *24*, 22–33.

Article

The Permeability of Natural versus Anthropogenic Forest Edges Modulates the Abundance of Ground Beetles of Different Dispersal Power and Habitat Affinity

Tibor Magura [1,*] and Gábor L. Lövei [2]

[1] Department of Ecology, University of Debrecen, 4032 Debrecen, Hungary
[2] Flakkebjerg Research Centre, Department of Agroecology, Aarhus University, 4200 Slagelse, Denmark; gabor.lovei@agro.au.dk
* Correspondence: maguratibor@gmail.com

Received: 31 July 2020; Accepted: 19 August 2020; Published: 21 August 2020

Abstract: Forest edges are formed by natural or anthropogenic processes and their maintaining processes cause fundamentally different edge responses. We evaluated the published evidence on the effect of various edges on the abundance of ground beetles of different habitat affinity and dispersal power. Our results, based on 23 publications and 86 species, showed that natural forest edges were impenetrable for open-habitat species with high dispersal power, preventing their influx into the forest interiors, while forest specialist species of limited dispersal power penetrated and reached abundances comparable to those in forest interiors. Anthropogenic edges, maintained by continued disturbance were permeable by macropterous open-habitat species, allowing them to invade the forest interiors, while such edges (except the forestry-induced ones) deterred brachypterous forest specialists. Different permeability of forest edges with various maintaining processes can affect ecosystem functions and services, therefore the preservation and restoration of natural forest edges are key issues in both forest ecology and nature conservation.

Keywords: anthropogenic edges; dispersal; edge effect; filter function; forest interior; forest specialist species; invasion; open-habitat species; natural edges; spillover

---

1. Introduction

With the ongoing, world-wide habitat conversion, previously continuous habitats become fragmented, and the presence of edges becomes more and more prevalent in landscapes [1]. Edges are transitional zones between different habitats that occur naturally, and have substantial influence on abiotic and biotic landscape conditions [2].

The boundary zone between adjacent habitats often display attributes that are distinct from either its adjacent habitat [3,4]. These special conditions create habitat for edge-inhabiting species with highly dynamic abundance and flow-on changes in their biotic interactions from predation to seed dispersal at these edges [3]. With edges becoming ubiquitous, they are one of the most studied entities in ecology [1].

Edge research, through examining manifold species and diverse types of edges has given us an articulate picture of the edge effects [5]. Four fundamental phenomena were identified to account for changes in species abundance across habitat edges: ecological flows, access to spatially separated resources, resource tracking and species interactions [5]. A predictive model, driven by resource distribution, aimed to forecast changes in abundance near edges for any species in any landscape [6]. Although several responses are indeed predictable, several unexplained responses limit the predictive power of this theory [6]. Edge orientation (edge position relative to the sun [5]), size and isolation of habitat patches, the quality of adjacent habitats, landscape composition [7], edge contrast (low vs. high [5,8]), the contrast between habitat patch and matrix [4], species traits (including habitat specialization, dispersal power, seasonal and diurnal activity, body size, reproduction [8,9], habitat suitability [8] and temporal effects from time of day to year [5]) can be responsible for the unexplained variation.

The edge-maintaining processes are also important drivers of the edge effect [10]. Magura et al.'s [11] "history-based edge effect hypothesis", assumes that natural vs. continued human influence via forestry, agriculture, or urbanization will impact the diversity of edge-associated assemblages. Ground beetles in forest edges maintained by natural processes are significantly more species rich than those in their interiors, while such difference does not exist at edges with continued human influence [11]. Species richness, however, could be an imperfect indicator of the edge effect, because of species-specific responses to the same stimuli [12–14]. Therefore, taking into account species traits during analysis could provide a deeper understanding. Here we evaluated the edge effect considering dispersal power (a life history trait) and habitat affinity (an ecological trait) of ground beetles at forest edges. These traits were selected because dispersal power is closely related to population turnover [15] affecting assemblage stability, while the selected, widely different habitat preferences are expected to generate deviations in response to habitat fragmentation [11]. Our hypotheses were: (1) for open-habitat ground beetle species of high dispersal power, forest edges maintained by natural processes constitute a barrier and prevent their influx into the forest interior, while (2) edges under continued human influences are penetrable; (3) for forest specialist carabids of limited dispersal power, edges maintained by natural processes are penetrable, while (4) edges with continued human influences are not. To test these hypotheses, we evaluated published evidence to compare the abundance of these groups at edges vs. interiors.

In the present study, we found support for all four hypotheses: different permeability of forest edges with various maintaining processes fundamentally determined the spatial dispersal of ground beetles with different dispersal power and habitat affinity.

2. Materials and Methods

2.1. Data Search and Selection

We searched the literature for relevant data on 27 May 2016 at the University of Debrecen, Hungary using the Web of Science platform with the "All databases" option. The following search string was used: TOPIC = (forest*) AND TOPIC = (edge* OR margin*) AND TOPIC = (carabid*), with a time period limit of 1975–2015. We also reviewed the bibliography of the papers found by the search for additional, relevant publications that had remained undetected. Our inclusion criteria were: the paper had to report data on carabid abundance, variability and sample sizes, from both forest interior and forest edge. From papers that studied carabids along transects, only data from the interior most location in the forest were used.

We found 204 relevant publications, 199 from the search in Web of Science and five from the reference lists of these papers. Of these, 53 papers reported abundance data from both forest interior and forest edge. Mean abundance of ground beetle species (with standard deviations and sample sizes) for forest interiors and edges were extracted from 23 studies (Table S1). Twelve papers studied forest edges with continued human influences, and 11 papers that studied forest edges maintained by natural processes. Of the edges with human influence, 5 were created by agriculture, 4 by forestry, and 3 by urbanization. Out of the five papers dealing with edges disturbed by agriculture, only one studied grassland as adjacent habitat, while the other four had agricultural fields as neighboring habitats, therefore these edges were grouped to "edges disturbed by agriculture". The 23 papers reported abundance data on 35 open-habitat species that were good dispersers ("flying grasslanders") and 51 poor disperser forest specialists ("walking specialists"; Table S2). Overall, our meta-analyses were based on 200 discrete edge-to-interior comparisons of abundance data concerning 86 ground beetle species (56 comparisons for open-habitat species and 144 ones for forest specialist species).

2.2. Classification of Edges Based on Their Maintaining Process

We classified forest edges according to their maintaining processes. In order to be classified as maintained by natural processes (succession), neighboring habitats (the forest interior and the adjacent grassland or meadows) had to be unmanaged (without cutting, thinning, intensive grazing, mowing or fire damage) for at least 50 years. Disturbance-maintained edges included those created by forestry (clear-cutting, or forest management operations), urbanization (forest patches embedded in, and adjacent to an urbanized area) or agriculture (the habitat neighboring the forest edge was cultivated, intensively grazed, mowed and/or regularly burned). We excluded edges where there was a mixture of various forces, and also those with shifts between natural and human influence over time.

2.3. Data Analyses

Ground beetles were categorized according to their dispersal power and habitat affinity. Short-winged (brachypterous) species were considered poor dispersers, while macropterous species were classified as good dispersers. Species associated with open-habitat were considered open-habitat species, while species restricted to forests were classified forest specialists. This categorization was made using information in the original papers; when this information was lacking, we consulted an online ground beetle database [16].

For each comparison, Hedge's unbiased standardized mean difference (Hedges' g) was calculated as:

$$g = J\frac{\overline{X_F} - \overline{X_E}}{S_{within}}, \tag{1}$$

$$S_{within} = \sqrt{\frac{(n_F - 1)S_F^2 + (n_E - 1)S_E^2}{n_F + n_E - 2}} \tag{2}$$

and

$$J = 1 - \frac{3}{4(n_F + n_E - 2) - 1} \tag{3}$$

where $\overline{X_F}$ and $\overline{X_E}$ denote the mean abundance of beetles in forest interior and forest edge, respectively, n_F and n_E are the sample sizes at the forest interior and forest edge, andS_F and S_E are their respective SDs. A negative g value indicates higher beetle abundance in forest edges than interiors, while a positive one shows higher abundance in forest interiors compared to forest edges.

Subgroup meta-analysis was used to examine whether the forest edge maintenance class (natural or anthropogenic) had an effect on ground beetle abundance. The overall effect and the effects of moderators (type of edge-maintaining process; type of human influence) were examined by random-effects models because of differences in geography, experimental conditions, design and research methods. One publication could provide data for several edge-to-interior abundance comparisons, thus we included a publication-level random effect as a nesting factor into the model. The mean effect size was considered statistically significant when the 95% bootstrap confidence interval (calculated from 999 iterations) did not include zero.

To describe heterogeneity, complementary measures of Q and I^2 were calculated [17]. Total variance (Q_{total}) was partitioned into within-(Q_{within}) and between group ($Q_{between}$) components and were tested for statistical significance [17]. Significant variance between groups ($Q_{between}$) means that edge effect on abundance significantly differed according to the edge-maintaining processes. During the calculations, only datasets with at least five edge-to-interior comparisons of abundance data from at least three different papers were included to keep statistical power. Publication bias was tested using funnel plots and the Egger test [17]. In case of significant asymmetry, the trim and fill method was employed [18]. Calculations were performed using the *MAd* (version 0.8-2 [19]) and *metafor* packages (version 1.9-9 [20]) in R programming environment (version 3.6.3 [21]).

3. Results

Considering all edges, the abundance of flying grasslanders (good disperser open-habitat species) was significantly higher in the edges than the interior (Figure 1a). Abundance pattern, however, was different according to the history of edges, although the between group variance ($Q_{between}$) was not significant (Table S3). At edges maintained by natural processes, the abundance of flying grasslanders was higher than those in the respective forest interiors, while no similar pattern occurred in edges maintained by human influence. Similarly, edges created by forestry activities showed no such difference (Figure 1a, Table S3). Neither the total nor the unexplained heterogeneity was significant (Table S3), and no significant funnel plot asymmetry was detected by the Egger tests (weighted or mixed-effects meta-regression) (Table S4). Nevertheless, the trim and fill method estimated 12 missing abundance data on the right side of the funnel plot (Figure S1a). Adding these data, however, did not change the significance of the overall effect in the model (Table S5).

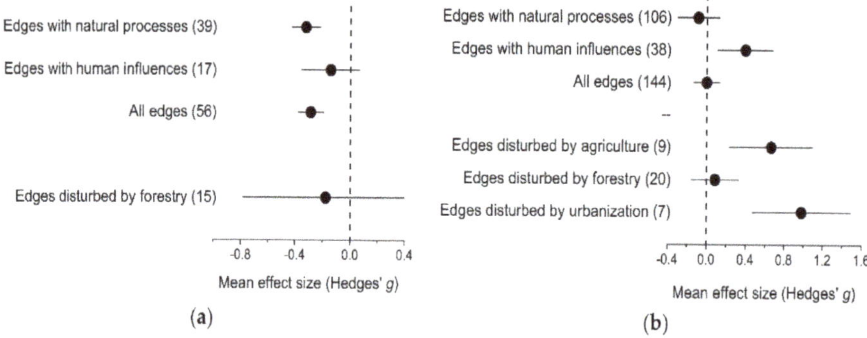

Figure 1. Mean effect sizes of random-effect models (± 95% confidence interval) for the abundance of flying grasslander (good disperser open-habitat) (**a**), and walking specialist (poor disperser forest specialist) ground beetle species (**b**). Values in brackets show the number of abundance values for which the mean effect size was calculated. Mean effect size was only calculated for edge-to-interior comparisons of abundance data with ≥5 cases. A negative *g* value indicates higher abundance in forest edges than interiors. Mean effect size is statistically significant when the confidence interval did not include zero.

Analyzing all edges together, there was no significant difference in the abundance of walking specialists (poor disperser forest specialists) between forest edges and their interiors (Figure 1b). The trends, however, was significantly different according to the maintaining forces (significant $Q_{between}$, Table S3). At edges with natural processes, the abundance of walking specialists was similar in edges and interiors, while in the case of edges with human influences, there were significantly fewer individuals there than in the forest interiors. Furthermore, the edge effect on the abundance of walking specialists was also significantly related to the type of human disturbance (significant $Q_{between}$, Table S3). In edges disturbed by agriculture or urbanization, the abundance at edges was significantly lower than in interiors, while there was no such difference at forestry-influenced edges (Figure 1b, Table S3). The total and the unexplained heterogeneities were significant in all models except in the case of agriculture-generated edges (Table S3). The Egger tests indicated significant funnel plot asymmetry (Table S4). The trim and fill method also estimated 7 missing abundance data on the left side (Figure S1b), but adding these did not change the non-significance of the overall effect (Table S5).

4. Discussion

The history-based edge effect hypothesis [11] assumes that different edge-maintaining processes (natural vs. continued human influence) have significant impacts on the spatio-temporal distribution of species across edges and the diversity of assemblages in edges and adjacent habitats. Indeed, ground beetle species richness was significantly higher in forest edges maintained by natural processes than the forest interiors, while there was no similar difference between edges with continued human influences and their respective forest interiors [11]. Species richness or taxonomic diversity, however, is not necessarily the most useful indicator to identify edge effects because species with various traits (in size, dispersal power, feeding habits, tolerance limits) may trigger different responses to changes in abiotic and biotic factors along edges [12–15]. Therefore, to reveal real ecological patterns, species with different traits should be analyzed separately to assess their responses to edge effects at both species and assemblage levels [22,23].

To date, the history-based edge effect hypothesis was scrutinized using body size [24], feeding habit [25], and habitat affinity [11] across variously maintained forest edges. The abundance of small-sized species was significantly higher at edges maintained by natural processes than in their interior, but no such difference was found in the case of forest edges maintained by agriculture or forestry. Furthermore, there is no significant difference in the abundance of either the medium-sized or the large-sized species between any type of forest edge and its interior [24]. Regarding feeding habits, the abundance of herbivorous, omnivorous, and predatory ground beetle species is significantly higher at edges with natural processes than their interiors, while no similar pattern occurs in edges with continued human influences [25]. The abundance of forest specialist species is not significantly different at edges maintained by natural processes with respect to their interiors. However, forest specialists avoid edges disturbed by agriculture or urbanization, while there is no such reaction to forestry-influenced edges [11]. Both the generalist and the open-habitat species are significantly more abundant in the edges than in interiors, irrespective of the kind of edge encountered. This difference is not registered when the edge was forestry-influenced [11].

Our study, combining a life history (wing morphology) and an ecological trait (habitat affinity), shows that the abundance of open-habitat species, which are good dispersers, was significantly higher in the edges maintained by natural processes compared to their interiors, while anthropogenic interventions, mainly forestry makes the edge penetrable, and their abundance becomes more even. Contrary to this, the forest specialist species, which are poor dispersers, approached and entered forest edges with natural processes and reached abundances similar to interiors, but they avoided edges influenced by agriculture or urbanization, although not forestry-disturbed edges. In accordance with previous results based on only a single ecological trait (habitat affinity), it seems that the filter function of edges is fundamentally different depending on their maintaining processes, that is, on their history [11]. For flying grasslander species inhabiting the surrounding open-habitats, edges with human influences, mainly by forestry activities, were penetrable, and these species also invaded the forest interior. However, naturally maintained forest edges, became impenetrable barriers, preventing the influx of these species into the forest interior. For walking specialist species, edges maintained by natural processes are penetrable, allowing them to disperse from forest interiors to edges, to move right even through the edges into the adjacent habitats [14]. Forestry-influenced edges seems to have filter function similar to edges with natural processes. However, edges maintained by agriculture or urbanization seemed impenetrable, preventing the dispersal of walking specialists into the edges, and thus limiting their possibility to disperse beyond the forest fragment. Different filtering function of forest edges with various maintaining processes was confirmed not only for ground beetles but also for other organisms and habitat parameters. Evaluating vegetation responses at boreal forest edges with various history, Harper et al. [26] also showed notable differences between forest structure responses to natural (fire) and anthropogenic (cut) edge influences.

Graduality (or abruptness) and permanence could be the main causes of the different filter function of variously maintained forest edges [26,27]. Natural processes (succession after natural disturbances, such as fire, insect outbreaks, windthrow, grazing and habitat disturbance by wildlife) create and maintain complex, heterogeneous, permanent, stratified, and gradual transitional zones extending up to 5–30 m outside the forest toward the adjacent habitat, and up to 10–30 m toward the interior of the forest [26,28–30]. These features of forest edges with natural processes constitute gradual changes in habitat structure and environmental conditions across the transitional zone. Permanent, successively changing structure and conditions, on the one hand, allow the spreading of walking specialist species from the forest interior to the edge, and provide an opportunity to disperse beyond the forest. On the other hand, these edges have a buffer function by preventing the influx of flying grasslander species into the forest interior [11]. Contrary to this, repeated anthropogenic disturbance (cultivation, intensive grazing, mowing, burning, forestry interventions, urban management operations) regularly destroys the development of complex, permanent, gradual edges, thus forming and maintaining a simplified, abrupt and often narrow transition zone. These characteristics of edges with human influences create abrupt changes in habitat structure and environmental parameters, limiting the permeability of these edges for forest specialist species, but from the other direction, allow the influx of open-habitat species into the forest interior [11].

The invasion of open-habitat species into the forest interiors may have harmful effects on the forest specialist species through biotic interactions, including resource competition and depletion, and even intraguild-predation [31]. Both intensive agriculture, forestry and progressive urbanization tend to eliminate natural, semi-natural habitat patches, including forest remnants, thus these human activities can be considered one of the greatest current threats to forest specialist species [32–38], impoverishing community composition and simplifying its organization [39,40], and damaging ecosystem functions and services [41,42]. The impoverishment and/or compositional changes of forest interior assemblages caused by the invasion of open-habitat species across edges with continued human influences, as well as the restricted dispersal or spillover of forest specialist species into these edges and the adjacent fields may have negative effects on ecosystem functions and services, like biological pest control, and decomposition [25]. Indeed, a recent study indicated limited predator spillover from native forest fragments across edges disturbed by agriculture to maize fields in central Argentina [43].

Our results also underline that forestry-induced edges are penetrable for walking specialist species, allowing them to disperse from forest interiors to edges. This result seems surprising, as it was shown that changes in habitat structure and environmental conditions in forest edges under forestry influences are detrimental for biodiversity [3]. Forestry-induced edges are mainly created by timber harvesting and the harvested sites will usually be reforested. Regeneration of these sites can reduce the contrast between the neighboring habitats, softening the edge effects [44]. Studying the edge influence on forest and understory structure and composition at forest edges adjacent to regenerating clear-cut originated sites, it was shown that significant edge effects are of relatively short duration [45]. In the first two years after harvesting, significant responses were detected to edge creation, but the edge influence weakens with time. This weakening of the edge influence resulted from the re-establishment of edge-related microclimatic gradients due to rapid regeneration of the adjacent, harvested habitat [45]. In forestry-generated edges, the regeneration of edge gradients in habitat structure and environmental conditions allow forest species to disperse into such edges. Moreover, this dispersion is strongly facilitated by the recovery potential of forest ground beetles after the canopy closure (8–16 years after the reforestation) in regenerating habitats [46–48].

Significant total and unexplained heterogeneity in the models concerning walking specialist species suggests that in addition to the history of forest edges, other features may be important in determining the spatial distribution of these species across edges. The size, isolation and quality of the neighboring habitats, the temporal effects and edge orientation are among important factors [5]. Moreover, other traits of the forest specialist species (body size, feeding habit, activity, and reproduction type) may also be responsible for the remaining heterogeneity, thus could be additional, important factors determining edge responses by these species. A global meta-analysis considering all the above-factors would be challenging but very useful to further examine and articulate the history-based edge effect.

Our results highlight that different maintaining processes (natural vs. anthropogenic) fundamentally determined the permeability of forest edges, the spatial dispersal of flying grasslander species and walking forest specialist ground beetle species across edges. This difference can basically affect both the biodiversity in edges and the local ecosystem functions and services. Therefore, all edges maintained by natural processes should be preserved and unfavorable changes to their structure and characteristics should be avoided to ensure their proper functioning. Simultaneously, if possible, human-induced edges should be restored (e.g., by softening theses edges [44]) to develop a filter function similar to edges with natural processes.

Supplementary Materials: The following are available online at http://www.mdpi.com/1424-2818/12/9/320/s1, Figure S1: Funnel plots with missing data (empty circles) for the abundance of flying grasslander (macropterous open-habitat) (a) and walking specialist (brachypterous forest specialist) ground beetle species (b), Table S1: Studies used in the meta-analyses, Table S2: Ground beetle species included into the meta-analyses, their dispersal power and habitat affinity, and the papers from which their abundances were extracted, Table S3: Estimates and heterogeneities in the models, Table S4: Results of regression test for funnel plot asymmetry of abundances of flying grasslander (macropterous open-habitat) and walking specialist (brachypterous forest specialist) ground beetle species, Table S5: Model results after trim and fill for the abundances of flying grasslander (macropterous open-habitat) and walking specialist (brachypterous forest specialist) ground beetle species.

Author Contributions: Conceptualization, T.M. and G.L.L.; methodology, T.M. and G.L.L.; validation, T.M. and G.L.L.; formal analysis, T.M.; investigation, T.M.; resources, T.M. and G.L.L.; data curation, T.M.; writing—original draft preparation, T.M. and G.L.L.; writing—review and editing, T.M. and G.L.L.; visualization, T.M.; funding acquisition, T.M. All authors have read and agreed to the published version of the manuscript.

Funding: This research was funded by the National Research, Development and Innovation Fund, grant number OTKA K-131459.

Acknowledgments: The authors would like to thank Béla Tóthmérész for discussions on the topic.

Conflicts of Interest: The authors declare no conflict of interest. The funder had no role in the design of the study; in the collection, analyses, or interpretation of data; in the writing of the manuscript, or in the decision to publish the results.

References

1. Fahrig, L. Ecological responses to habitat fragmentation per se. *Annu. Rev. Ecol. Evol. Syst.* **2017**, *48*, 1–23. [CrossRef]
2. Turner, M.G.; Gardner, R.H. *Landscape Ecology in Theory and Practice. Pattern and Process*, 2nd ed.; Springer: New York City, NY, USA, 2015.
3. Murcia, C. Edge effects in fragmented forests: Implications for conservation. *Trends Ecol. Evol.* **1995**, *10*, 58–62. [CrossRef]
4. Ewers, R.M.; Didham, R.K. Continuous response functions for quantifying the strength of edge effects. *J. Appl. Ecol.* **2006**, *43*, 527–536. [CrossRef]
5. Ries, L.; Fletcher, R.J.; Battin, J.; Sisk, T.D. Ecological responses to habitat edges: Mechanisms, models, and variability explained. *Annu. Rev. Ecol. Evol. Syst.* **2004**, *35*, 491–522. [CrossRef]
6. Ries, L.; Sisk, T.D. A predictive model of edge effects. *Ecology* **2004**, *85*, 2917–2926. [CrossRef]
7. Hardt, E.; Pereira-Silva, E.F.L.; Dos Santos, R.F.; Tamashiro, J.Y.; Ragazzi, S.; Lins, D.B.D.S. The influence of natural and anthropogenic landscapes on edge effects. *Landsc. Urban. Plan.* **2013**, *120*, 59–69. [CrossRef]
8. Peyras, M.; Vespa, N.I.; Bellocq, M.I.; Zurita, G.A. Quantifying edge effects: The role of habitat contrast and species specialization. *J. Insect Conserv.* **2013**, *17*, 807–820. [CrossRef]
9. Carvajal-Cogollo, J.E.; Urbina-Cardona, N. Ecological grouping and edge effects in tropical dry forest: Reptile-microenvironment relationships. *Biodivers. Conserv.* **2015**, *24*, 1109–1130. [CrossRef]
10. Strayer, D.L.; Power, M.E.; Fagan, W.F.; Pickett, S.T.A.; Belnap, J. A classification of ecological boundaries. *Bioscience* **2003**, *53*, 723–729. [CrossRef]
11. Magura, T.; Lövei, G.L.; Tóthmérész, B. Edge responses are different in edges under natural versus anthropogenic influence: A meta-analysis using ground beetles. *Ecol. Evol.* **2017**, *7*, 1009–1017. [CrossRef]

12. Koivula, M.; Hyyryläinen, V.; Soininen, E. Carabid beetles (Coleoptera: Carabidae) at forest-farmland edges in southern Finland. *J. Insect Conserv.* **2004**, *8*, 297–309. [CrossRef]
13. Brigić, A.; Starčević, M.; Hrašovec, B.; Elek, Z. Old forest edges may promote the distribution of forest species in carabid assemblages (Coleoptera: Carabidae) in Croatian forests. *Eur. J. Entomol.* **2014**, *111*, 715–725. [CrossRef]
14. Magura, T. Ignoring functional and phylogenetic features masks the edge influence on ground beetle diversity across forest-grassland gradient. *For. Ecol. Manag.* **2017**, *384*, 371–377. [CrossRef]
15. den Boer, P.J. *Dispersal Power and Survival: Carabids in a Cultivated Countryside*, 1st ed.; H. Veenman & Zonen, B. V.: Wageningen, The Netherlands, 1977.
16. Homburg, K.; Homburg, N.; Schäfer, F.; Schuldt, A.; Assmann, T. Carabids.org–A dynamic online database of ground beetle species traits (Coleoptera, Carabidae). *Insect Conserv. Divers.* **2014**, *7*, 195–205. [CrossRef]
17. Borenstein, M.; Hedges, L.V.; Higgins, J.P.T.; Rothstein, H.R. *Introduction to Meta-Analysis*, 1st ed.; John Wiley & Sons Ltd.: Chichester, UK, 2009.
18. Duval, S.; Tweedie, R. Trim and fill: A simple funnel-plot-based method of testing and adjusting for publication bias in meta-analysis. *Biometrics* **2000**, *56*, 455–463. [CrossRef]
19. Del Re, A.C.; Hoyt, W.T. MAd: Meta-Analysis with Mean Differences. Available online: https://cran.r-project.org/web/packages/MAd/MAd.pdf (accessed on 14 February 2020).
20. Viechtbauer, W. Conducting meta-analyses in R with the metafor package. *J. Stat. Softw.* **2010**, *36*, 1–48. [CrossRef]
21. R Core Team. R: A Language and Environment for Statistical Computing. R Foundation for Statistical Computing: Vienna, Austria, 2017. Available online: http://www.R-project.org/ (accessed on 14 February 2020).
22. Evans, M.J.; Banks, S.C.; Davies, K.F.; Mcclenahan, J.; Melbourne, B.; Driscoll, D.A. The use of traits to interpret responses to large scale—Edge effects: A study of epigaeic beetle assemblages across a Eucalyptus forest and pine plantation edge. *Landsc. Ecol.* **2016**, *31*, 1815–1831. [CrossRef]
23. Magura, T.; Lövei, G.L. Environmental filtering is the main assembly rule of ground beetles in the forest and its edge but not in the adjacent grassland. *Insect Sci.* **2018**, *26*, 154–163. [CrossRef]
24. Magura, T.; Lövei, G.L. The type of forest edge governs the spatial distribution of different-sized ground beetles. *Acta Zool. Acad. Sci. Hung.* **2020**, submitted.
25. Magura, T.; Lövei, G.L.; Tóthmérész, B. Various edge response of ground beetles in edges under natural versus anthropogenic influence: A meta-analysis using life-history traits. *Acta Zool. Acad. Sci. Hung.* **2019**, *65*, 3–20. [CrossRef]
26. Harper, K.A.; Macdonald, S.E.; Mayerhofer, M.S.; Biswas, S.R.; Esseen, P.-A.; Hylander, K.; Stewart, K.J.; Mallik, A.U.; Drapeau, P.; Jonsson, B.-G.; et al. Edge influence on vegetation at natural and anthropogenic edges of boreal forests in Canada and Fennoscandia. *J. Ecol.* **2015**, *103*, 550–562. [CrossRef]
27. Bowersox, M.A.; Brown, D.G. Measuring the abruptness of patchy ecotones: A simulation-based comparison of landscape pattern statistics. *Plant. Ecol.* **2001**, *156*, 89–103. [CrossRef]
28. Larrivée, M.; Drapeau, P.; Fahrig, L. Edge effects created by wildfire and clear-cutting on boreal forest ground-dwelling spiders. *For. Ecol. Manag.* **2008**, *255*, 1434–1445. [CrossRef]
29. Magura, T. Carabids and forest edge: Spatial pattern and edge effect. *For. Ecol. Manag.* **2002**, *157*, 23–37. [CrossRef]
30. Harper, K.A.; Drapeau, P.; Lesieur, D.; Bergeron, Y. Forest structure and composition at fire edges of different ages: Evidence of persistent structural features on the landscape. *For. Ecol. Manag.* **2014**, *314*, 131–140. [CrossRef]
31. Tscharntke, T.; Tylianakis, J.M.; Rand, T.A.; Didham, R.K.; Fahrig, L.; Batáry, P.; Bengtsson, J.; Clough, Y.; Crist, T.O.; Dormann, C.F.; et al. Landscape moderation of biodiversity patterns and processes—Eight hypotheses. *Biol. Rev.* **2012**, *87*, 661–685. [CrossRef]
32. Tilman, D.; Fargione, J.; Wolff, B.; D'Antonio, C.; Dobson, A.; Howarth, R.; Schindler, D.; Schlesinger, W.H.; Simberloff, D.; Swackhamer, D. Forecasting agriculturally driven global environmental change. *Science* **2001**, *292*, 281–284. [CrossRef]
33. Magura, T.; Lövei, G.L.; Tóthmérész, B. Does urbanization decrease diversity in ground beetle (Carabidae) assemblages? *Glob. Ecol. Biogeogr.* **2010**, *19*, 16–26. [CrossRef]
34. Kromp, B. Carabid beetles in sustainable agriculture: A review on pest control efficacy, cultivation impacts and enhancement. *Agric. Ecosyst. Environ.* **1999**, *74*, 187–228. [CrossRef]

35. Fenoglio, M.S.; Rossetti, M.R.; Videla, M. Negative effects of urbanization on terrestrial arthropod communities: A meta-analysis. *Glob. Ecol. Biogeogr.* **2020**, *29*, 1412–1429. [CrossRef]
36. Lövei, G.L.; Magura, T. Ground beetle (Coleoptera: Carabidae) diversity is higher in narrow hedges composed of a native compared to non-native trees in a Danish agricultural landscape. *Insect Conserv. Divers.* **2017**, *10*, 141–150. [CrossRef]
37. Magura, T.; Ferrante, M.; Lövei, G.L. Only habitat specialists become smaller with advancing urbanisation. *Glob. Ecol. Biogeogr.* **2020**, *29*, in. [CrossRef]
38. Paillet, Y.; Bergès, L.; Hjältén, J.; Ódor, P.; Avon, C.; Bernhardt-Römermann, M.; Bijlsma, R.-J.; De Bruyn, L.; Fuhr, M.; Grandin, U.; et al. Biodiversity differences between managed and unmanaged forests: Meta-analysis of species richness in Europe. *Conserv. Biol.* **2010**, *24*, 101–112. [CrossRef]
39. Magura, T.; Lövei, G.L.; Tóthmérész, B. Conversion from environmental filtering to randomness as assembly rule of ground beetle assemblages along an urbanization gradient. *Sci. Rep.* **2018**, *8*, 16992. [CrossRef] [PubMed]
40. Gayer, C.; Lövei, G.L.; Magura, T.; Dieterich, M.; Batáry, P. Carabid functional diversity is enhanced by conventional flowering fields, organic winter cereals and edge habitats. *Agric. Ecosyst. Environ.* **2019**, *284*, 106579. [CrossRef]
41. Eötvös, C.B.; Magura, T.; Lövei, G.L. A meta-analysis indicates reduced predation pressure with increasing urbanization. *Landsc. Urban. Plan.* **2018**, *180*, 54–59. [CrossRef]
42. Eötvös, C.B.; Lövei, G.L.; Magura, T. Predation pressure on sentinel insect prey along a riverside urbanization gradient in Hungary. *Insects* **2020**, *11*, 97. [CrossRef]
43. Ferrante, M.; González, E.; Lövei, G.L. Predators do not spill over from forest fragments to maize fields in a landscape mosaic in central Argentina. *Ecol. Evol.* **2017**, *7*, 7699–7707. [CrossRef]
44. Samways, M.J. Insect conservation: A synthetic management approach. *Annu. Rev. Entomol.* **2007**, *52*, 465–487. [CrossRef]
45. Harper, K.A.; Macdonald, S.E. Structure and composition of edges next to regenerating clear-cuts in mixed-wood boreal forest. *J. Veg. Sci.* **2002**, *13*, 535–546. [CrossRef]
46. Magura, T.; Bogyó, D.; Mizser, S.; Nagy, D.D.; Tóthmérész, B. Recovery of ground-dwelling assemblages during reforestation with native oak depends on the mobility and feeding habits of the species. *For. Ecol. Manag.* **2015**, *339*, 117–126. [CrossRef]
47. Pawson, S.M.; Brockerhoff, E.G.; Watt, M.S.; Didham, R.K. Maximising biodiversity in plantation forests: Insights from long-term changes in clearfell-sensitive beetles in a Pinus radiata plantation. *Biol. Conserv.* **2011**, *144*, 2842–2850. [CrossRef]
48. Jung, J.-K.; Lee, J.-H. Trait-specific responses of carabid beetle diversity and composition in Pinus densiflora forests compared to broad-leaved deciduous forests in a temperate region. *Diversity* **2020**, *12*, 275. [CrossRef]

© 2020 by the authors. Licensee MDPI, Basel, Switzerland. This article is an open access article distributed under the terms and conditions of the Creative Commons Attribution (CC BY) license (http://creativecommons.org/licenses/by/4.0/).

Article

Trait-Specific Responses of Carabid Beetle Diversity and Composition in *Pinus densiflora* Forests Compared to Broad-Leaved Deciduous Forests in a Temperate Region

Jong-Kook Jung [1,*,†] and Joon-Ho Lee [1,2]

1 Entomology Program, Department of Agricultural Biotechnology, Seoul National University, Gwanakro 1, Gwanakgu, Seoul 08826, Korea; jh7lee@snu.ac.kr

2 Research Institute of Agriculture and Life Sciences, Seoul National University, Gwanakro 1, Gwanakgu, Seoul 08826, Korea

* Correspondence: jk82811@korea.kr; Tel.: +82-2-961-2665; Fax: +82-2-961-2679

† Present address: Division of Forest Insect Pests and Diseases, National Institute of Forest Science, Seoul 02455, Korea.

Received: 16 June 2020; Accepted: 7 July 2020; Published: 9 July 2020

Abstract: Since successful reforestation after the 1970s, Korean red pine (*Pinus densiflora*) forests have become the most important coniferous forests in Korea. However, the scarcity of evidence for biodiversity responses hinders understanding of the conservation value of Korean red pine forests. This study was conducted to explore the patterns of carabid beetle diversity and assemblage structures between broad-leaved deciduous forests and *P. densiflora* forests in the temperate region of central Korea. Carabid beetles were sampled by pitfall trapping from 2013 to 2014. A total of 66 species were identified from 9541 carabid beetles. Species richness in broad-leaved deciduous forests was significantly higher than that in pine forests. In addition, the species composition of carabid beetles in broad-leaved deciduous forests differed from that of *P. densiflora* forests. More endemic, brachypterous, forest specialists, and carnivorous species were distributed in broad-leaved deciduous forests than in *P. densiflora* forests. Consequently, carabid beetle assemblages in central Korea are distinctively divided by forest type based on ecological and biological traits (e.g., endemisim, habitat types, wing forms, and feeding guilds). However, possible variation of the response of beetle communities to the growth of *P. densiflora* forests needs to be considered for forest management based on biodiversity conservation in temperate regions, because conifer plantations in this study are still young, i.e., approximately 30–40-years old.

Keywords: biodiversity conservation; temperate forests; ground beetles; ecological trait

---

1. Introduction

The positive ecological role in plantation forests has been emphasized recently because it prevents the loss of biodiversity caused by deforestation worldwide [1,2]. For example, plantations can have direct impacts on biodiversity [3], as well as stand dynamics and structure [4]. The global plantation area is approximately 7.3% of the total forested area (291 million ha) [5]. However, conifer forests in Korea, mainly plantations, cover approximately 36.9% of the total forested area (2.3 million ha) [6]. Almost all natural forests in Korea had been destroyed until the 1960s and have recovered since the 1970s [7]. For this reason, a large area of forests in Korea is composed of 30–50-year-old conifer plantations and naturally regenerating deciduous forests covering approximately 87.3% of the total forested area [6]. Thus, understanding biotic responses of young forests, including plantations and regenerated forests, provides valuable insight into biodiversity conservation in temperate forests of Korea.

Among young conifer plantation forests in Korea, *Pinus densiflora* forests are the most important, covering approximately 24.7% (1,562,843 ha) of the total forested area [6]. However, biodiversity in *P. densiflora* forests, as compared with young broad-leaved deciduous forests (*Quercus* spp.) in Korea, is poorly understood despite successful reforestation since the 1970s. In human-dominated landscapes, the diversity and composition of carabid beetles [8–11] and moths [12] in coniferous forests (mainly *P. densiflora*) have been investigated and compared to those in secondary mixed forests. In mountains in temperate regions, the effects of forest types on moths [13,14] and carabids [15–17] have been compared. Differences in carabid communities among forest types have been occasionally observed due to the effects of habitat fragmentation [8,11] and landscape heterogeneity [10]. However, the species diversity and composition of moths [13,14] and carabids [9,15–17] in coniferous plantations are generally similar to those in secondary or natural forests, except that the species richness of all carabid beetles is increased in regenerating deciduous forests [10]. In temperate regions, few studies have considered the ecological and biological traits of carabid beetles to compare diversity between forest types, such as habitat type [10,11], body size [10], and wing morphs [10,11,17]. By considering ecological and biological traits, the species richness of macropterous species differed only in grasslands as compared with various forest types, such as two natural forests (broad-leaved deciduous and *P. densiflora* forests) and one plantation (a deciduous coniferous forest) [17], and the species richness of brachypterous and forest specialists decreased significantly as a result of habitat fragmentation irrespective of forest type [11]. Nonetheless, the low number of spatial replications for measuring biodiversity in young *P. densiflora* forests is limited to emphasizing the conservation value compared to young broad-leaved deciduous forests because the distribution of insects is basically related to the local environment, such as habitat heterogeneity and elevation, in addition to habitat types.

To compare biodiversity in young *P. densiflora* and young broad-leaved deciduous forests, we studied carabid beetles because they are diverse, ecologically well known, and abundant in most ecosystems [18]. In particular, large-bodied and flightless carabid beetles are more vulnerable to disturbances than generalist species that have high mobility [19]. The endemism rate could also be an important biological characteristic because endemism is closely related to the low dispersal ability of carabid beetles combined with adaptation to specific local environments [20]. In addition, feeding guilds could also be an important trait, because diversity and distribution of carnivorous and herbivorous species are influenced by vegetation [21] and microhabitat characteristics, such as leaf litter [22]. These ecological and biological traits could provide a basis for biodiversity conservation in different forest types.

The primary aim of this study was to compare the abundance, species richness, and composition of carabid beetles based on ecological and biological traits (all carabid beetles, geographical distribution, wing morph, habitat preference, and feeding guilds) between young broad-leaved deciduous forests (*Quercus* spp. dominated, abbreviated to BLF hereafter) and young pine plantations (*P. densiflora* dominated, abbreviated to PF hereafter) in central Korea. In addition, we characterized carabid communities in each forest type by indicator species analysis, because carabid communities can not be regenerated even in relatively old plantations [23].

2. Materials and Methods

2.1. Study Area

Carabid beetles were studied in Yeongwol-gun and Jeongseon-gun, which are located in central Korea (Figure 1a). Because of the higher proportion of mountainous forests in the study area with steep terrain, forest landscapes are well preserved. In Yeongwol-gun and Jeongseon-gun, forests cover approximately 85% of the area [24]. The study area is surrounded by several reserves, such as Mt. Chiaksan National Park (17,567 ha), Mt. Sobaeksan National Park (32,205 ha), Mt. Taebaeksan

National Park (7005 ha), and forest genetic resource reserves on Mt. Gariwangsan (2475 ha) (Figure 1b). Moreover, the Donggang River traversing Yeongwol-gun, Jeongseon-gun, and Pyeongchang-gun has been designated an ecological and landscape conservation area (6497 ha) because of its beautiful landscape and high biodiversity. The climate of the region is temperate, with an average annual temperature of 11.03 °C; the average temperatures of the warmest and coolest months from 2011 to 2013 were 31.07 °C and −11.77 °C, respectively; the annual precipitation in 2011, 2012, and 2013 was 2085.7, 1398.8, and 1245.5 mm, respectively [25].

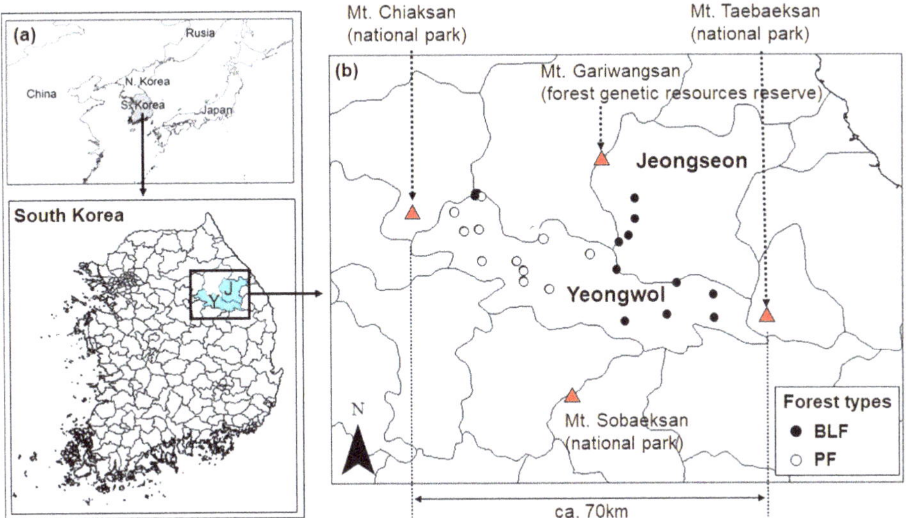

Figure 1. Map of (**a**) study area and (**b**) study sites (filled circle, broad-leaved deciduous forests (BLF); opened circle, *Pinus densiflora* forests (PF)).

Carabid beetles were sampled at 22 study sites (Figure 1b). The location and environmental characteristics of the study sites are described in Table 1. Half of the 22 sites were PF, the other half BLF. Forest information (forest types and ages) in this study area was confirmed by the Korean forest geographic information service system [6]. The latitude and longitude (i.e., spatial location) of the study sites were 37°05′03″ N, 37°22′23″ N and 128°13′48″ E, 128°48′35″ E, respectively, but the spatial locations of the two forest types were quite distinct (Figure 1b). In addition, forest ages of PF were relatively higher than those of BLF (Welch two sample t-test, $t = 3.49$, $p = 0.002$). However, elevation was not different between the two forest types (range of 245–473 m for PF and 273–744 m for BLF, $t = 1.49$, $p = 0.157$) (Table 1). These five variables for each sampling site were used in the redundancy analysis (RDA) to explain the variation in carabid beetle assemblages between forest types.

2.2. Sampling

Three pitfall traps were installed at each study site for collecting carabid beetles from 2013 to 2014. Sampling periods differed among study sites, ranging between 78–196 days (234–588 trap × days, Table 1). Short sampling periods at some study sites, such as YNC, YNG, and YBD, may have limited appropriate carabid beetle collection. However, we sampled carabid beetles at least from August to October at all sampling sites because the diversity of carabid beetles during these months in mountainous forests of central Korea was higher than that in other months [26].

Table 1. Environmental characteristics for each study site with sampling information.

Location		Code for Site	Forest Type [a]	Forest Age Class [b]	Latitude	Longitude	Elevation (m)	Trap × Days	Sampling Period
Jeongseon-eup	Gwangha-ri	DJGw	BLF	III	37°21′39″ N	128°37′43″ E	296	330	Jul–Oct 2013
	Gasu-ri	DJGa	BLF	III	37°18′49″ N	128°37′41″ E	285	330	Jul–Oct 2013
	Unchi-ri	DJU	BLF	III	37°16′36″ N	128°36′48″ E	278	330	Jul–Oct 2013
	Deokcheon-ri	DJD	BLF	III	37°15′43″ N	128°35′23″ E	273	330	Jul–Oct 2013
Yeongwol-eup	Geoun-ri	DYG	PF	V	37°14′09″ N	128°31′11″ E	258	330	Jul–Oct 2013
Suju-meyon	Beopheung-ri	YSB1	PF	IV	37°22′18″ N	128°15′45″ E	473	330	Jul–Oct 2013
	Beopheung-ri	YSB2	BLF	IV	37°22′23″ N	128°15′24″ E	488	588	Apr–Oct 2014
	Unhak-ri	YSU	PF	V	37°19′53″ N	128°12′24″ E	461	579	Jul 2013–Jun 2014
	Mureung-ri	YSM	PF	V	37°17′28″ N	128°15′50″ E	280	579	Jul 2013–Jun 2014
Jucheon-myeon	Docheon-ri	YJD	PF	V	37°17′16″ N	128°13′48″ E	323	579	Jul 2013–Jun 2014
	Yongseok-ri	YJYo	PF	IV	37°13′12″ N	128°16′14″ E	355	579	Jul 2013–Jun 2014
Hanbando-myeon	Hutan-ri	YHH	PF	IV	37°11′56″ N	128°22′03″ E	245	579	Jul 2013–Jun 2014
	Ongjeong-ri	YHO	PF	IV	37°13′07″ N	128°21′08″ E	303	579	Jul 2013–Jun 2014
Jungdong-myeon	Jikdong-ri	YJJ	BLF	IV	37°10′13″ N	128°43′27″ E	434	339	Jun–Oct 2014
	Yeonsang-ri	YJYe	BLF	IV	37°12′01″ N	128°35′07″ E	332	339	Jun–Oct 2014
Sangdong-eup	Naedeok-ri	YSN	BLF	IV	37°08′37″ N	128°48′32″ E	664	339	Jun–Oct 2014
	Deokgu-ri	YSD	BLF	IV	37°05′29″ N	128°48′35″ E	744	339	Jun–Oct 2014
Gimsatgat-myeon	Nae-ri	YGN	BLF	IV	37°05′53″ N	128°42′03″ E	525	339	Jun–Oct 2014
	Waseok-ri	YGW	BLF	III	37°05′03″ N	128°36′14″ E	339	339	Jun–Oct 2014
Nam-myeon	Gwangcheon-ri	YNG	PF	V	37°09′21″ N	128°25′38″ E	266	234	Aug–Oct 2014
	Changwon-ri	YNC	PF	V	37°10′31″ N	128°22′06″ E	298	234	Aug–Oct 2014
Buk-myeon	Deoksang-ri	YBD	PF	III	37°17′19″ N	128°23′43″ E	350	234	Aug–Oct 2014

[a] Abbreviation of habitat types are: BLF, broad-leaved deciduous forests; PF, *Pinus densiflora* dominated forest; [b] Forest age classes in our study sties were confirmed by the forest geographic information service system in Korea (Korea Forest Service, 2016): III, 21–30-years old; IV, 31–40-years old; V, 41–50-years old.

The trap was a plastic cup (9.5 cm in diameter, 10 cm in height, 430 mL in volume), and a plastic roof was placed 3 cm above each trap to prevent the inflow of rainfall and litter. Pitfall traps were unbaited, containing preservatives (200 mL, 95% ethyl-alcohol/95% ethylene-glycol = 1:1) as killing-preserving solutions, which were replaced every 4 weeks. Collected carabid beetles were transported to the laboratory, dried, mounted, and identified to the species level under a dissecting microscope (63×, Olympus SZ61, Tokyo, Japan). Identification was performed according to [27–35]. Nomenclature confirmed the list of Korean Carabidae by [32–35]. Endemism with respect to Korea, including South and North Korea, was confirmed according to [32,34,36]. The identified carabid beetles were stored in the Insect Ecology Laboratory at Seoul National University.

2.3. *Data Analysis*

All of the following statistical approaches were performed in R version 3.3.2 [37]. The species richness of carabid beetles in each forest type was estimated by individual-based rarefaction curves using the "vegan" R package [38]. Rarefaction curves are based on a random resampling of the pool of captured individuals and are used to estimate expected richness at lower sample sizes [39]. Rarefaction methods enable meaningful standardization and comparison of datasets [39]. For the rarefaction curves, carabid beetle samples were pooled based on forest type and forest age.

To compare the abundance and species richness of total carabid beetles and functional groups between two forest types, a t-test was applied. For the t-test, carabid beetle samples were pooled at each site. In addition, we standardized species richness and abundance, dividing them by trap × days (ranging between 234 and 588 trap × days) for statistical analyses (Table 1) because sampling periods were different at each study site and some pitfall traps were disturbed by rain.

The RDA was performed to explain the variation in carabid beetle assemblages between forest types and to visualize differences in species composition between forest types. Because there is a possible "horseshoe effect" in ordination [40], a Hellinger transformation was applied to species data prior to RDA. To preselect the environmental variables, we applied the "ordistep" function from the "vegan" R package [41], which performs both forward and backward selection of variables based on *P*-values. The vegan function "anova" was used to assess the significance of each variable in the final model (sequential test) and to obtain the *P*-values for each variable based on permutation tests [42]. We further compared the species richness of different ecological groups of carabid beetles by performing a Pearson correlation analysis against plot scores of the first and second axes resulting from the ordination analysis (plot scores for RDA1 and RDA2 based on weighted averages of site scores).

The indicator value (IndVal) approach was conducted to find indicator species between forest types that could characterize the habitats [43]. Flexible IndVal was independent of the relative abundance of other species, and there was no need to use pseudospecies. IndVal is at maximum value (1.00) when all individuals of a species occur in a single group of sites and when the species is found in all sites of that group. Therefore, abundance and occurrence stability indices for species were determined for analysis. The statistical significance of the species indicator value ($\alpha = 0.1$) was determined using the Monte Carlo permutation test [44,45].

3. Results

3.1. *Diversity and Abundance of Carabid Beetles*

A total of 66 species were identified from 9541 collected carabid beetles (Table S1). Three *Synuchus* species, i.e., *Synuchus nitidus* (3888 individuals, 40.75% of total), *Synuchus cycloderus* (2587 individuals, 27.11%), and *Synuchus agonus* (825 individuals, 8.65%), and *Eucarabus cartereti*, 1982 (418 individuals, 4.38%), were abundant, comprising over 80% of the total carabid beetle assemblages (Table S1). Fourteen endemic specieswere collected, comprising 8.84% of the total abundance (843 individuals).

Individual-based rarefaction curves indicated that the species richness of carabid beetles in BLF was higher than that in PF (Figure 2a). When considering forest age for each forest type, the species richness of carabid beetles in 31–40-year-old BLF was more distinct than the other three forest types (Figure 2b). Considering ecological groups, the abundances of brachypterous, forest specialists, open-habitat species, and carnivorous species were not different between the two forest types, while those of widespread, macropterous, and herbivorous species were significantly higher in PF (Table 2). The abundance of endemic species was higher only in BLF. The species richness of endemic, brachypterous, forest specialist, and carnivorous species, in PF were significantly lower than those in BLF.

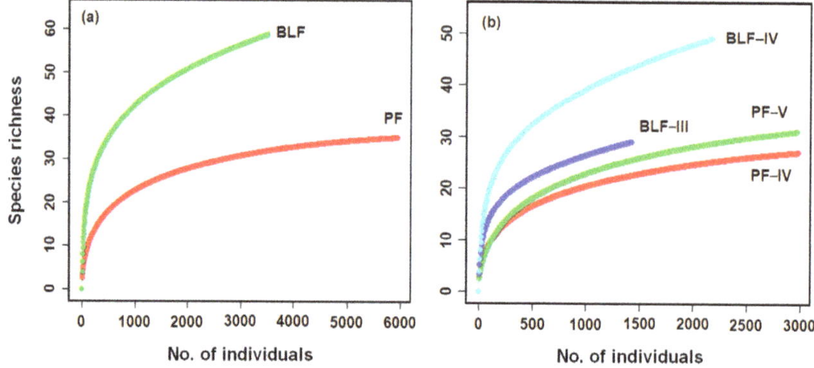

Figure 2. Individual-based rarefaction curves for ground-beetle catches (**a**) in *Pinus densiflora* forests (PF) and broad-leaved deciduous forests (BLF), and (**b**) in forest age classes (III, 21–30-years old; IV, 31–40-years old; V, 41–50-years old). Data points indicate average species numbers computed for the given number of individuals.

Table 2. Comparison of carabid catches (abundance and species richness) between *Pinus densiflora* forests (PF) and broad-leaved deciduous forests (BLF). Bold characters indicate statistical difference of abundance or species richness of carabid beetles between BLF and PF ($p < 0.05$).

Dependent Variables	Carabid Catches (Mean ± S.E.)		Statistics	
	PF (n = 11)	BLF (n = 11)	t-value	p
Abundance				
All species	540.6 ± 96.46	326.7 ± 77.63	1.728	0.1002
Endemic species	**7.6 ± 2.47**	**69.0 ± 27.40**	**−2.231**	**0.0494**
Widespread species	**533.0 ± 95.44**	**257.7 ± 56.44**	**2.483**	**0.0243**
Brachypterous species	53.3 ± 14.87	172.0 ± 70.11	−1.657	0.1261
Macropterous species	**487.4 ± 87.95**	**154.7 ± 33.31**	**3.537**	**0.0037**
Forest specialists	519.0 ± 91.69	301.5 ± 79.25	1.795	0.0881
Open-habitat species	21.6 ± 12.47	25.3 ± 8.67	−0.239	0.8135
Carnivorous species	539.5 ± 96.24	318.4 ± 75.34	0.942	0.3573
Herbivorous species	**1.2 ± 0.38**	**8.4 ± 2.76**	**−3.002**	**0.0084**
Species richness				
All species	**10.0 ± 0.88**	**15.4 ± 1.99**	**−2.466**	**0.0274**
Endemic species	**1.1 ± 0.31**	**4.5 ± 0.90**	**−3.534**	**0.0039**
Widespread species	8.9 ± 0.72	10.9 ± 1.32	−1.327	0.2038
Brachypterous species	**4.5 ± 0.58**	**8.0 ± 1.41**	**−2.330**	**0.0362**
Macropterous species	5.5 ± 0.78	7.4 ± 1.13	−1.325	0.2021
Forest specialists	**7.2 ± 0.60**	**11.3 ± 1.71**	**−2.262**	**0.0423**
Open-habitat species	2.8 ± 0.72	4.1 ± 0.99	−1.041	0.3116
Carnivorous species	**9.2 ± 0.77**	**13.3 ± 1.75**	**−2.459**	**0.0265**
Herbivorous species	0.82 ± 0.26	2.1 ± 0.55	−0.9464	0.3629

3.2. Species Composition of Carabid Beetles

The RDA for carabid beetles was significant (permutation test for RDA under reduced model, $F_{4,17} = 3.51, p < 0.001$). The first and second axes accounted for 36.55% of the total variation, with the first and second axes explaining 24.17% (permutation test, $F_{1,17} = 7.51, p < 0.001$) and 12.37% ($F_{1,17} = 3.84$, $p = 0.005$), respectively (Table 3). The first axis of the RDA was strongly and positively associated with elevation and longitude, while forest type was negatively related to the first RDA axis (Figure 3 and Table 3). In addition, the species composition in BLF appeared to be distinct from that in PF (Figure 3). Species richness in most ecological groups showed positive correlations with weighted average site scores on the first RDA axis, whereas species richness of open-habitat species showed weak and positive correlations with the first axis (Table 3).

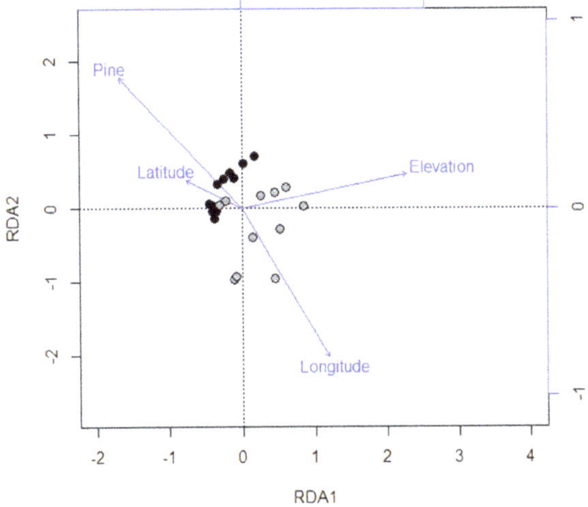

Figure 3. A RDA ordination of species composition of ground beetles in 22 sites. The first and second RDA axes are shown. The different forest types are shown as closed circles by different colors, i.e., black, broad-leaved deciduous forests and grey, *Pinus densiflora* forests.

Table 3. Correlations between environmental variables or species richness and the axes in the RDA analyses (total inertia = 0.3808). Species richness was calculated by biological or ecological groups.

Variables	RDA 1	RDA 2	RDA 3	RDA 4
Eigenvalue	0.0921	0.0471	0.0218	0.0114
% variation explain †	24.17	12.37	5.72	3.00
Environmental variables				
Forest type	** −0.64	*** 0.54	−0.02	−0.23
Elevation	*** 0.87	0.14	0.05	−0.32
Latitude	−0.29	0.12	*** 0.76	0.23
Longitude	* 0.45	** −0.62	−0.20	−0.25
Species richness				
Total	*** 0.85	−0.18	0.06	−0.16
Endemic	*** 0.86	−0.07	0.08	−0.14
Widespread	*** 0.80	−0.23	0.06	−0.17
Brachypterous	*** 0.79	−0.06	0.09	−0.20
Macropterous	* 0.43	−0.32	−0.02	−0.05
Forest specialists	*** 0.83	−0.01	0.02	−0.18
Open-habitat	0.24	* −0.43	0.08	−0.04
Carnivorous species	*** 0.76	−0.11	0.04	−0.20
Herbivorous species	** 0.55	* −0.48	0.07	0.00

† Percentage = 100 × (variation explained by respective axis)/(variation explained by all environmental variables). Statistically significant correlations are indicated by * $p < 0.05$, ** $p < 0.01$, *** $p < 0.001$.

According to forest types, the following two characteristic groups of carabid beetle species were detected: (1) numerous in BLF, i.e., *Eucarabus cartereti cartereti*, *Harpalus discrepans*, *Pristosiavigil*, *Leistus niger*, and *Cymindis collaris*; (2) abundant in PF, i.e., *Coptolabrus smaragdinus branickii* (Table 4). However, *P. vigil*, *C. collaris*, and *C. s. branickii* had low indicator values ($p > 0.05$).

Table 4. Two-way indicator table showing carabid beetle species indicator value for habitat clustering hierarchy according to forest type, the number of individuals of ground beetle species by forest types, and their proportion in observation (%). Carabid species that showed high indicator value ($p < 0.1$) were listed.

Forest Types	Indicator Value	*p*-Value	Number of Individuals (Proportion in Observation, %)	
			QF	PF
Broad-leaved deciduous forest (QF)				
Eucarabus cartereti cartereti	0.998	0.001	416 (100.0)	2 (9.1)
Harpalus discrepans	0.749	0.023	30 (63.6)	4 (27.3)
Pristosia vigil	0.729	0.060	77 (54.5)	2 (18.2)
Leistus niger niger	0.674	0.033	8 (45.5)	
Cymindis collaris	0.603	0.093	8 (36.4)	
Pinus densiflora-dominated forest (PF)				
Coptolabrus smaragdinus branickii	0.699	0.081	6 (27.3)	51 (54.5)

4. Discussion

4.1. Carabid Beetle Diversity in Different Forest Types

In temperate forests of Korea composed of young coniferous and deciduous forests, we found that forest type and elevation were important factors that influenced carabid beetle assemblages. In particular, the species richness of carabid beetles of the BLF in this study was obviously higher than inPF, although the average elevation did not differ between the forest types. These results were rather different from those in previous studies conducted in Korea [15], China [16], and Japan [17], whereas similar findings were reported in Japan [46] and Korea [10]. In particular, the diversity of carabid beetles was significantly reduced in conifer plantations, especially in more heterogeneous landscapes [10]. In Japan, carabid beetles in fragmented landscapes at high elevations (1240–1490 m) showed no differences in diversity between natural forests (evergreen coniferous and broad-leaved deciduous) and plantations (deciduous coniferous) [17]. In contrast, similar carabid diversity between pine plantations and oak forests was reported from Korea and China [15,16], but the estimated species richness in mixed forests was higher than that reported in China [16]. On the basis of ecological traits of carabid beetles in several habitat types, including deciduous, mixed, and mature conifer forests, more forest specialists were caught in deciduous forests, while more forest generalists were caught in mature conifer plantations [46]. Thus, conifer plantations sometimes support biodiversity, but this is dependent on the history of disturbances such as forest management or biogeographical characteristics. In fact, many forest-inhabiting species in Korea are flightless, and thus they cannot disperse from source habitats to fragments, especially the large-bodied species [47]. However, our study region is composed of continuous mountainous forests, not fragmented landscapes, although the locations of the study sites of each forest type were not pairwise in the same locality. Thus, PF could have the potential to support low biodiversity, at least for carabid beetles.

Although this study showed that the diversity of carabid beetles clearly differed between forest types, the synergistic effect of elevation, locality of study sites, and forest type could also be important to understand carabid beetle assemblages in this region. Thus, interpretation of our results should be carefully applied to forest management for biodiversity conservation. This is because the difference in carabid communities could be due to the mixed effect of forest types and other environmental variables, such as moisture, pH, organic matter, texture [17,48], leaf litter [22], canopy cover [16], and elevation [49].

4.2. Ecological and Biological Traits of Carabid Beetles and Forest Types

Plant species (i.e., forest type) are regarded as the primary factor for insect communities, especially herbivorous insects [50,51]. However, understanding the distributional pattern of carnivorous carabids in temperate forests could be more important as compared with phytophagous or omnivorous carabids because, in general, the biodiversity of carnivorous carabids is habitat heterogeneity dependent [52]. In fact, most carabid beetle species in this study were carnivorous (i.e., 9436 individuals belonging to 54 species), whereas only 105 belonging to 12 species were phytophagous or omnivorous (e.g., species belonging to genus *Amara* and *Harpalus*), which could occasionally occur in forests with understory vegetation because of forest gaps or forest roads. Thus, temperate forests in Korea could have no herbivorous carabid beetles. Nonetheless, forest type can alter other herbivorous insect communities, and carnivorous carabid beetles could change according to changes in the abundance and composition of herbivorous communities.

Unlike feeding guilds, this study suggested that the trait-specific response of carabid beetles to different forest types appears to be valuable information for establishing biodiversity conservation plans in young reforested landscapes. In this study, the species richness of endemic, brachypterous, and forest specialists in BLF was obviously higher than that in PF, while the abundance of macropterous and widespread species in PF was obviously higher than that in BLF. In general, highly heterogeneous topography due to mountains is regarded as one of the reasons for the high diversity of forest specialist carabid beetles in Korea. In fact, many forest specialists are brachypterous and they prefer stable environments, such as BLF in this study. PF with homogenized environments can be suitable habitats for macropterous and widespread species, such as *S. cycloderus* and *S. nitidus*. In fact, these two species, which are dominant in forests of South Korea [10], were significantly more common in PF than in BLF and accounted for approximately 68% of all carabid beetles in this study. Consequently, ecological and biological traits for carabid beetles are very useful to understand the distributional pattern of biodiversity in temperate forests and are beneficial for forestry through biodiversity conservation.

4.3. Habitat Specialists for Forest Types

This study showed that only three species (*E. cartereti*, *H. discrepans*, and *L. niger*) could be considered BLF specialists, although the species composition was quite different between forest types. However, some species or groups could have the potential to become bioindicators of distinct forest types. For example, species, belonging to the genus *Pterostichus* were more numerous in BLF (258 individuals belonging to 8 species) than in PF (69 individuals belonging to three species) (Table S1).

In contrast, there were no habitat specialists in PF, only *C. s. branickii* was more frequently observed in PF than in BLF. *C. s. branickii* is a habitat specialist in PF patches [11], and this species is also found in open habitats, such as agricultural fields, orchards, and lawns [53,54]. Thus, this species appears to be a habitat generalist. In the genus *Synuchus*, some species, such as *S. cycloderus* and *S. nitidus*, can also be used as potential bioindicators. Although they were found in almost every study site, these two species were abundant in PF. This could be largely due to the habitat preference of *Synuchus* towards dry forests [55]. In fact, *P. densiflora* trees are generally planted on south-facing slopes of mountains in Korea, and the trees prefer well-drained soils. In addition, *P. densiflora* is considered to be the most preferable tree species for plantations in Korea because of its esthetic value [56]. Thus, habitat conditions in PF are generally dry, especially in fragmented patches [11]. For these reasons, some *Synuchus* species, especially *S. cycloderus* and *S. nitidus*, are potential bioindicators in PF. Nonetheless, many PF in this study were still young, approximately 31–50-years old. Therefore, habitat specialists did not have enough time to establish populations in PF.

5. Conclusions

This study showed that carabid beetle assemblages in temperate forests of central Korea were distinctively divided by forest type based on ecological and biological traits (e.g., endemism, habitat types, and wing forms). This suggested that monoculture plantations in temperate regions (i.e., *P. densiflora* in this study), which appeared to be simple habitats, could have a limited ability to preserve high biodiversity, at least for carabid beetles. In particular, young *P. densiflora* plantations may not be appropriate for supporting populations of endemic, brachypterous, and forest specialist species. For biodiversity conservation in Korea as a reforested area, however, possible variation of the beetle community response to forest growth of *P. densiflora* plantations need to be considered.

Supplementary Materials: The following are available online at http://www.mdpi.com/1424-2818/12/7/275/s1, Table S1: List of carabid beetles with number of individuals in *Pinus densiflora* forests (PF) and broad-leaved deciduous forests (BLF) in central Korea.

Author Contributions: Conceptualization, J.-K.J. and J.-H.L.; Data curation, J.-K.J.; Formal analysis, J.-K.J.; Funding acquisition, J.-H.L.; Investigation, J.-K.J.; Methodology, J.-K.J.; Project administration, J.-H.L.; Visualization, J.-K.J.; Writing—original draft, J.-K.J.; Writing—review and editing, J.-H.L. All authors have read and agreed to the published version of the manuscript.

Funding: This research was funded by Brain Korea 21 plus.

Conflicts of Interest: The authors declare no conflict of interest.

References

1. Brockerhoff, E.G.; Jactel, H.; Parrotta, J.A.; Quine, C.P.; Sayer, J. Plantation forests and biodiversity: Oxymoron or opportunity? *Biodivers. Conserv.* **2008**, *17*, 925–951. [CrossRef]
2. Hartmann, H.; Daoust, G.; Bigué, B. Negative or positive effects of plantation and intensive forestry on biodiversity: A matter of scale and perspective. *For. Chron.* **2010**, *86*, 354–364. [CrossRef]
3. Lindenmayer, D.B.; Hobbs, R.J.; Salt, D. Plantation forests and biodiversity conservation. *Aust. For.* **2003**, *66*, 62–66. [CrossRef]
4. Carnus, J.-M.; Parrotta, J.; Brockerhoff, E.G.; Arbez, M.; Jactel, H.; Kremer, A.; Lamb, D.; O'Hara, K.; Walters, B. Planted forests and biodiversity. *J. For. Ecol.* **2006**, *104*, 65–77.
5. MacDicken, K.; Jonsson, Ö.; Piña, L.; Maulo, S.; Adikari, Y.; Garzuglia, M.; Lindquist, E.; Reams, G.; D'Annunzio, R. *The Global Forest Resources Assessment 2015: How Are the World's Forests Changing?* Food and Agriculture Organization of the United Nations: Rome, Italy, 2015.
6. Korea Forest Service. Statistical Yearbook of Forestry No. 46. Korea Forest Service, 2016. Available online: https://www.forest.go.kr (accessed on 6 April 2017).
7. Lee, D.K. *Ecological Management of Forests*; Seoul National University Press: Seoul, Korea, 2012.
8. Do, Y. The effect of fragmentation and intensive management on carabid beetles in coniferous forest. *Appl. Ecol. Environ. Res.* **2013**, *11*, 451–461. [CrossRef]
9. Jung, J.-K.; Kim, S.-T.; Lee, S.-Y.; Yoo, J.-S.; Lee, J.-H. Comparison of Ground Beetle Communities (Coleoptera: Carabidae) between Coniferous and Deciduous Forests in Agricultural Landscapes. *J. For. Environ. Sci.* **2013**, *29*, 211–218. [CrossRef]
10. Jung, J.-K.; Kim, S.-T.; Lee, S.Y.; Park, C.-G.; Park, J.-K.; Lee, J.-H. A comparison of diversity and species composition of ground beetles (Coleoptera: Carabidae) between conifer plantations and regenerating forests in Korea. *Ecol. Res.* **2014**, *29*, 877–887. [CrossRef]
11. Jung, J.-K.; Lee, S.K. Trait-specific response of ground beetles (Coleoptera: Carabidae) to forest fragmentation in the temperate region in Korea. *Biodivers. Conserv.* **2018**, *27*, 53–68. [CrossRef]
12. Choi, S.W.; Park, M.; Kim, H. Differences in moth diversity in two types of forest patches in an agricultural landscape in Southern Korea—Effects of habitat heterogeneity. *J. Ecol. Field. Biol.* **2009**, *32*, 183–189. [CrossRef]
13. Choi, S.-W. Diversity and composition of larger moths in three different forest types of Southern Korea. *Ecol. Res.* **2008**, *23*, 503–509. [CrossRef]
14. Yi, H.-B.; Kim, H.-J. Species Composition and Species Diversity of Moths (Lepidoptera) on Quercus mongolica forests sand Pinus densiflora forests, in Korean National Long-term Ecological Research Sites (Mt. Nam, Mt. Jiri, Mt. Wolak). *Korean J. Appl. Èntomol.* **2010**, *49*, 105–113. [CrossRef]
15. Chang, S.J.; Kim, J.K. Distribution of carabid beetles (Coleoptera: Carabidae) in different forests of central Kangwon-do. *J. For. Sci. Kangwon Nat. Univ.* **2000**, *16*, 42–49.

16. Warren-Thomas, E.; Zou, Y.; Dong, L.; Yao, X.; Yang, M.; Zhang, X.; Qin, Y.; Liu, Y.; Sang, W.; Axmacher, J.C. Ground beetle assemblages in Beijing's new mountain forests. *For. Ecol. Manag.* **2014**, *334*, 369–376. [CrossRef]
17. Ogai, T.; Kenta, T. The effects of vegetation types and microhabitats on carabid beetle community composition in cool temperate Japan. *Ecol. Res.* **2016**, *31*, 177–188. [CrossRef]
18. Lövei, G.L.; Sunderland, K.D. Ecology and Behavior of Ground Beetles (Coleoptera: Carabidae). *Annu. Rev. Èntomol.* **1996**, *41*, 231–256. [CrossRef] [PubMed]
19. Rainio, J.; Niemelä, J. Ground beetles (Coleoptera: Carabidae) as bioindicators. *Biodivers. Conserv.* **2003**, *12*, 487–506. [CrossRef]
20. Schuldt, A.; Assmann, T. Environmental and historical effects on richness and endemism patterns of carabid beetles in the western Palaearctic. *Ecography* **2009**, *32*, 705–714. [CrossRef]
21. Harvey, J.A.; Van Der Putten, W.H.; Turin, H.; Wagenaar, R.; Bezemer, T.M. Effects of changes in plant species richness and community traits on carabid assemblages and feeding guilds. *Agric. Ecosyst. Environ.* **2008**, *127*, 100–106. [CrossRef]
22. Koivula, M.; Punttila, P.; Haila, Y.; Niemelä, J. Leaf litter and the small-scale distribution of carabid beetles (Coleoptera, Carabidae) in the boreal forest. *Ecography* **1999**, *22*, 424–435. [CrossRef]
23. Elek, Z.; Magura, T.; Tóthmérész, B. Impacts of non-native Norway spruce plantation on abundance and species richness of ground beetles (Coleoptera: Carabidae). *Web Ecol.* **2001**, *2*, 32–37. [CrossRef]
24. KOSIS. Korean Statistical Information Service. 2015. Available online: http://kosis.kr/ (accessed on 8 March 2015).
25. KMA. Korea Meteorological Administration. 2016. Available online: http://www.kma.go.kr/ (accessed on 28 August 2016).
26. Jung, J.-K.; Suk, S.-W.; Kim, B.-Y.; Hong, E.J.; Kim, Y.; Jeong, J.-C. Differences in temporal variation of ground beetle assemblages (Coleoptera: Carabidae) between two well-preserved areas in Mt. Sobaeksan National Park. *J. For. Environ. Sci.* **2017**, *33*, 122–129.
27. Habu, A. *Fauna Japonica, Carabidae Truncatipennes Group (Insecta: Coleoptera)*; Biogeographical Society of Japan: Japan, Tokyo, 1967.
28. Habu, A. *Fauna Japonica, Carabidae: Harpalini (Insecta: Coleoptera)*; Keigaku Publishing Co. Ltd.: Tokyo, Japan, 1973.
29. Habu, A. *Fauna Japonica, Carabidae: Platynini (Insecta: Coleoptera)*; Keigaku Publishing Co. Ltd.: Tokyo, Japan, 1978.
30. Habu, A. Classification of the Callistini of Japan (Coleoptera, Carabidae). *Entomol. Rev. Jpn.* **1987**, *42*, 1–36.
31. Kwon, Y.J.; Lee, S.M. Classification of the subfamily Carabinae from Korea (Coleoptera: Carabidae). *Insecta Koreana* **1984**, *4*, 1–363.
32. Park, J.K. *Subfamily Carabinae in Korea (Coleoptera: Carabidae)*; Economic Insects of Korea 23; Junghaeng-Sa: Seoul, Korea, 2004.
33. Park, J.K.; Choi, I.J.; Park, J.; Choi, E.Y. *Insect Fauna of Korea, vol. 12, no. 16, Arthropoda: Insecta: Coleoptera: Carabidae: Chlaeniini, Truncatipennes Group: Odacanthinae, Lebiinae*; Junghaengsa, Inc.: Incheon, Korea, 2014.
34. Park, J.K.; Paik, J.C. *Family Carabidae. Economic Insects of Korea* 12; Junghaeng-Sa: Seoul,, Korea, 2001.
35. Park, J.K.; Park, J. *Insect Fauna of Korea, vol. 12, no. 13, Arthropoda: Insecta: Coleoptera: Carabidae: Pterostichinae*; Junghaengsa, Inc.: Incheon, Korea, 2014.
36. Löbl, I.; Smetana, A. *Catalogue of Palaearctic Coleoptera. Archostemata-Myxophaga-Adephaga*; Apollo Books: Stenstrup, Denmark, 2003; Volume 1.
37. R Core Team. *R: A Language and Environment for Statistical Computing*; R Foundation for Statistical Computing: Vienna, Austria. Available online: http://www.R-project.org/ (accessed on 6 April 2017).
38. Oksanen, J.; Blanchet, F.G.; Friendly, M.; Kindt, R.; Legendre, P.; McGlinn, D.; Minchin, P.R.; O'Hara, R.B.; Simpson, G.L.; Solymos, P.; et al. Vegan: Community Ecology Package. Available online: https://cran.r-project.org/web/packages/vegan/vegan.pdf (accessed on 1 May 2020).
39. Gotelli, N.J.; Colwell, R.K. Quantifying biodiversity: Procedures and pitfalls in the measurement and comparison of species richness. *Ecol. Lett.* **2001**, *4*, 379–391. [CrossRef]
40. Legendre, P.; Gallagher, E. Ecologically meaningful transformation for ordination of species data. *Oecologia* **2001**, *129*, 271–280. [CrossRef]

41. Oksanen, J. Multivariate Analysis of Ecological Communities in R: Vegan Tutorial. 2007. Available online: https://www.researchgate.net/publication/260136364_Multivariate_Analysis_of_Ecological_Communities_in_R_Vegan_Tutorial (accessed on 14 April 2017).
42. Legendre, P.; Oksanen, J.; Ter Braak, C.J.F. Testing the significance of canonical axes in redundancy analysis. *Methods Ecol. Evol.* **2010**, *2*, 269–277. [CrossRef]
43. Dufrêne, M.; Legendre, P. Species Assemblages and Indicator Species: The Need for a Flexible Asymmetrical Approach. *Ecol. Monogr.* **1997**, *67*, 345–366. [CrossRef]
44. Legendre, P.; Legendre, L. *Numerical Ecology*; Elsevier: Amsterdam, The Netherlands, 1998.
45. McGeoch, M.A.; Chown, S.L. Scaling up the value of bioindicators. *Trends Ecol. Evol.* **1998**, *13*, 46–47. [CrossRef]
46. Fuller, R.J.; Oliver, T.H.; Leather, S.R. Forest management effects on carabid beetle communities in coniferous and broadleaved forests: Implications for conservation. *Insect Conserv. Divers.* **2008**, *1*, 242–252. [CrossRef]
47. Kotze, D.J.; O'Hara, R.B. Species decline—But why? Explanation of carabid beetle (Coleoptera, Carabidae) declines in Europe. *Oecologia* **2003**, *135*, 138–148. [CrossRef]
48. Antvogel, H.; Bonn, A. Environmental parameters and micospatial distribution of insects: A case study of carabids in an alluvial forest. *Ecography* **2001**, *24*, 470–482. [CrossRef]
49. Eyre, M.D.; Rushton, S.P.; Luff, M.L.; Telfer, M.G. Investigating the relationships between the distribution of British ground beetle species (Coleoptera, Carabidae) and temperature, precipitation and altitude. *J. Biogeogr.* **2005**, *32*, 973–983. [CrossRef]
50. Usher, M.B.; Keiller, S.W.J. The macrolepidoptera of farm woodlands: Determinants of diversity and community structure. *Biodivers. Conserv.* **1998**, *7*, 725–748. [CrossRef]
51. Lepš, J.; Novotny, V.; Basset, Y. Habitat and successional status of plants in relation to the communities of their leaf-chewing herbivores in Papua New Guinea. *J. Ecol.* **2001**, *89*, 186–199. [CrossRef]
52. Purtauf, T.; Roschewitz, I.; Dauber, J.; Thies, C.; Tscharntke, T.; Wolters, V. Landscape context of organic and conventional farms: Influences on carabid beetle diversity. *Agric. Ecosyst. Environ.* **2005**, *108*, 165–174. [CrossRef]
53. Liu, Y.; Axmacher, J.C.; Wang, C.; Li, L.; Yu, Z. Ground beetles (Coleoptera: Carabidae) in the intensively cultivated agricultural landscape of northern China—Implications for biodiversity conservation. *Insect Conserv. Diver.* **2010**, *3*, 34–43. [CrossRef]
54. Jung, J.-K.; Lee, J.-H. Forest–farm edge effects on communities of ground beetles (Coleoptera: Carabidae) under different landscape structures. *Ecol. Res.* **2016**, *31*, 799–810. [CrossRef]
55. Fujita, A.; Maeto, K.; Kagawa, Y.; Itô, N. Effects of forest fragmentation on species richness and composition of ground beetles (Coleoptera: Carabidae and Brachinidae) in urban landscapes. *Èntomol. Sci.* **2008**, *11*, 39–48. [CrossRef]
56. Korea Forest Research Institute. Economic Tree Species 1: Pine Tree. Korea Forest Service, 2012. Available online: http://know.nifos.go.kr (accessed on 1 June 2020).

© 2020 by the authors. Licensee MDPI, Basel, Switzerland. This article is an open access article distributed under the terms and conditions of the Creative Commons Attribution (CC BY) license (http://creativecommons.org/licenses/by/4.0/).

MDPI

Article

Unequivocal Differences in Predation Pressure on Large Carabid Beetles between Forestry Treatments

Jana Růžičková * and Zoltán Elek

MTA-ELTE-MTM Ecology Research Group, Biological Institute, Eötvös Loránd University, Pázmány Péter Sétány 1/C, H-1117 Budapest, Hungary; zoltan.elek2@gmail.com
* Correspondence: jr.tracey@seznam.cz

Abstract: Carabid beetles (Coleoptera: Carabidae) are considered as one of the most cardinal invertebrate predatory groups in many ecosystems, including forests. Previous studies revealed that the predation pressure provided by carabids significantly regulates the ecological network of invertebrates. Nevertheless, there is no direct estimation of the predation risk on carabids, which can be an important proxy for the phenomenon called ecological trap. In our study, we aimed to explore the predation pressure on carabids using 3D-printed decoys installed in two types of forestry treatments, preparation cuts and clear cuts, and control plots in a Hungarian oak–hornbeam forest. We estimated the seasonal, diurnal and treatment-specific aspects of the predation pressure on carabids. Our results reveal a significantly higher predation risk on carabids in both forestry treatments than in the control. Moreover, it was also higher in the nighttime than daytime. Contrarily, no effects of season and microhabitat features were found. Based on these clues we assume that habitats modified by forestry practices may act as an ecological trap for carabids. Our findings contribute to a better understanding of how ecological interactions between species may change in a modified forest environment.

Keywords: 3D printing; artificial prey; behavior; clear cut; ecological trap; ground beetles; preparation cut; sentinel prey method

Citation: Růžičková, J.; Elek, Z. Unequivocal Differences in Predation Pressure on Large Carabid Beetles between Forestry Treatments. *Diversity* **2021**, *13*, 484. https://doi.org/10.3390/d13100484

Academic Editor: Tibor Magura

Received: 17 September 2021
Accepted: 1 October 2021
Published: 3 October 2021

Publisher's Note: MDPI stays neutral with regard to jurisdictional claims in published maps and institutional affiliations.

Copyright: © 2021 by the authors. Licensee MDPI, Basel, Switzerland. This article is an open access article distributed under the terms and conditions of the Creative Commons Attribution (CC BY) license (https://creativecommons.org/licenses/by/4.0/).

1. Introduction

Predation by various terrestrial animals can be one of the most important drivers of carabid evolution and, together with abiotic factors, determines their activity patterns [1]. Although carabids have various morphological and chemical defense mechanisms to avoid or reduce predation risk [2–4], they can also shift their behavior in space and time by being active only in a certain part of the day or using different habitat patches than predators. Another option is to be hidden in the soil, under the leaf litter or understory vegetation. For instance, studies describing the activity patterns of some large carabid species reported relatively long periods of beetle inactivity between movements that can last even for a couple of days [5,6]. In some individuals, regardless of sex, this no-movement behavior dominated during the tracking period [7]. Elek et al. [8] suggested that these periods when beetles are hidden and not moving could be a predator avoidance strategy.

Although carabids are often studied as important natural enemies [9–11], they can also be predated due to their middle position in the food chain. In temperate forests of the Northern Hemisphere, carabids represent potential prey for a large spectrum of predators, including bats, hedgehogs, shrews, raccoon dogs, wild boars and frogs [3,12–14]. Interactions between carabids and their predators might change due to shifts in the distribution of suitable habitat patches of various sizes within forest stands as a result of forestry practices [15–17]. This is especially true for Europe, where the majority of temperate forests have a semi-natural origin [18]. Therefore, it is crucial to explore whether the spatial differences created by forestry practices can affect the predation pressure on carabids. In addition, it can be an important issue to assess whether a certain treatment can act as an ecological trap for carabids [19], a habitat which is suitable for foraging or breeding, despite the fact that mortality can be higher there than in the unmanaged control forest stands.

There is an emerging need to estimate the structural changes in communities towards understanding ecosystem functions. Predation is well known as a key driver of these processes, potentially forming the community itself (i.e., keystone predation), and any change in predation intensity can reflect changes in ecosystem functions [20]. The technique involving artificial decoys as prey provides a powerful tool to assess relative rates of predation or predation pressure across various treatments [21]. Artificial prey is relatively cheap and easy to manipulate, and it is not invasive as it does not involve living specimens [22–24]. This can be an issue especially for rare or endangered species with low population densities where collection is nearly impossible [14]. Using 3D-printed decoys seems to be more suitable than involving dried fragile specimens that can be easily damaged [25]. Other frequently used materials, such as clay or plasticine [9,10,26], cannot replicate the narrow parts of beetles' body, such as legs and antennae [14]. Yet, these materials and techniques can be used for modeling various invertebrates with simple body shapes, such as snails, slugs, caterpillars and earthworms, and is called the sentinel prey method [9,27].

In this paper, we experimentally tested the predation pressure on carabids in the Hungarian managed temperate forest using real-sized 3D-printed models as decoys. As a model organism for 3D decoys, we selected a forest generalist, *Carabus coriaceus* L., 1758, a large species commonly occurring in Hungarian oak–hornbeam forests that sensitively reacts to forestry practices [8,28]. Two distinct types of forestry treatments, clear cuts and preparations cuts, were used for the field experiment. They considerably differ from surrounded unmanaged forest stands in terms of tree height and the cover of understory vegetation, leaf litter and bare soil [29] as well as in microclimatic conditions [30,31]. Hence, we expected that the predation pressure will vary between forestry treatments and undisturbed control stands due to the different availability of shelters formed by dense ground vegetation or leaf litter. Moreover, the different activity patterns of potential predators may presume that the predation pressure will also have diurnal and seasonal aspects based on predator activity and breeding period. Following these clues, we tested how the predation pressure on large carabids is affected at the different spatio-temporal scales. In particular, we focused on microhabitat characteristic, habitat type and time to determine the most influential factor(s) of the predation pressure in the studied forest stands.

2. Materials and Methods

2.1. Study Area

We conducted our experiment in the vicinity of Pilisszántó village in the Pilis Mountains, the northern part of Hungary (N 47°40′, E 18°54′). The study area (40 ha) is a structurally homogenous, 80-year-old, managed two-layered sessile oak–hornbeam forest. The structure of the stand is a result of the past and recent management under the shelterwood silvicultural system. The upper canopy layer is dominated by sessile oak (*Quercus petraea* (Matt.) Liebl, 1784) with an average height of 21 m and a mean diameter of 28 cm at breast height. The second most abundant tree species, hornbeam (*Carpinus betulus* L., 1753), forms a secondary canopy layer with an average height of 11 m and a mean diameter of 12 cm. Other admixing tree species include *Quercus cerris* L., 1753, *Fagus sylvatica* L., 1753, *Prunus avium* L., 1753 and *Fraxinus ornus* L., 1753. The shrub layer is scarce, mainly consisting of the regeneration of hornbeam and *F. ornus*, and the understory cover is dominated by mesic forest plants, such as *Carex pilosa* Scopoli, 1772 and *Melica uniflora* Retz, 1779 [16,29].

Our study was implemented as a part of the Pilis Forestry System Experiment [32] where four forestry treatments representing two different silvicultural systems were established in 2014. The main aim of this project is to explore the major effect of various treatments on natural forest regeneration and the biodiversity of several taxa including plants, enchytraeid worms, spiders and ground beetles (see Elek et al. [16] for further details). For the field experiment conducted in this study, we used two of the four implemented treatments representing characteristic stages of rotation forestry system: (1) Clear cut (CC) was a circular clear-cutting area of 80 m diameter surrounded by a closed-canopy stand. (2) Preparation cut (P) was created when 30% of the total basal area of the dominant tree layer and the whole secondary tree layer were removed in a spatially uniform way in a circle of 80 m diameter. These two treatments were chosen due to their strong effects on carabids at different levels, from community composition to individual activity [8,16,17].

2.2. Three-Dimensional Printed Decoys and Field Experiment

The 3D model of *Carabus coriaceus* was prepared in Blender 2.8 [33] based on high-resolution photos of real-sized individuals. One of the advantages of using *C. coriaceus* is its relatively large body size (33–40 mm), making the species suitable for detailed 3D printing. Unlike the other large species, *C. coriaceus* is unified in color (black) without iridescence and it is also one of the most abundant species in the area [17]. The beetle model (Supplementary Material Figure S1) was converted into printing data by KISSlicer 1.6, and then life-size three-dimensional decoys were generated by the 3D printer (DeltiQ M, developed by TriLAB Group s.r.o., Brno, Czech Republic). Black polylactic acid (PLA) filament was used as a production material. The printed decoy was the same size as the real specimen (Figure 1a).

Figure 1. Printed decoy of *Carabus coriaceus* in comparison with a real specimen (**a**). Decoys installed on the green cardboard (**b**) were checked for turns (**c**), broken or bent parts (**d**), predation attempts by birds (**e**), wild boars (**f**) or relocations (**g**).

The field experiment was conducted in two seasons, spring and autumn, which corresponded with the highest activity peaks of carabids in the area [16,17]. In both treatments (CC and P) and control plots (C), we installed 10 decoys (two lines, each with five beetles) on the ground with an A5 cardboard sheet beneath them. Each plot (both treatments and control) was replicated in three blocks, resulting in nine sampling plots. The distance between decoys was approximately 2.5 m. Decoys were positioned in the center of the green-colored sheet to avoid any unintended attraction due to conspicuous sheet colors (Figure 1b). The exact position was marked by two perpendicular lines drawn on the paper; the decoy was installed at their intersection to be able to record possible predation events. We recorded the cover of bare soil, litter, herbal and shrub layer (in %) in a one-meter-radius circle around each decoy and the number of surrounding trees (see summary Table 1). These environmental variables can potentially affect the distribution of carabids via the availability of shelters [7,8]. In total, 90 *C. coriaceus* decoys were installed per season. Decoys were checked twice a day, in the morning and evening, for six consecutive days (7–12 September 2020 and 8–13 June 2021). The exposed decoy was considered attacked when it had been turned or moved/relocated from its original position, including its disappearance or damage, such as scratches, bites and missing parts. For turns, we measured the angle; if relocated, the distance was recorded. Although decoy turns suggest that the potential predator was interested in the exposed decoy, the relocation or scratches are clear signs of a predation attempt. After measurements, decoys were repositioned or replaced by new ones when necessary.

Table 1. Environmental characteristics of forestry treatments represented as the cover of bare soil, leaf litter, herbal layer and shrubbery and the average number of trees per plot.

Treatment	Cover of (Mean ± SEM, in %)				No. of Trees per Plot (Mean)
	Bare Soil	**Leaf Litter**	**Herbal Layer**	**Shrubs**	
Control	2.51 ± 0.52	68.70 ± 2.68	28.25 ± 2.59	0.58 ± 0.37	4.67
Clear cut	3.63 ± 1.04	10.91 ± 1.68	36.10 ± 2.20	49.35 ± 2.90	0.50
Preparation cut	4.11 ± 0.97	33.36 ± 2.39	50.40 ± 2.76	12.10 ± 2.14	2.67

2.3. *Data Analyses*

The predation pressure was considered as a ratio between attack and no-attack events on 10 decoys in a plot per measurement session. This response was coded as two-column matrix [attack; no attack] using the *cbind* function in R 3.6.1 [34] where all analyses were conducted. We used generalized linear models (the *glm* function) with a binomial distribution and logit link function. Three different models were built considering various temporal and spatial scales. In the first model, treatment (factor with three levels: control, preparation cuts, clear cuts) was used as a single explanatory variable (spatial scale). In the second model, treatment, cover of leaf litter, bare soil and number of trees in the plot were included (micro-spatial scale). Cover of herbs and shrubs were excluded from the analyses due to a strong negative correlation with leaf litter (Pearson r = −0.48 for herb layer and r = −0.70 for shrubbery; see Supplementary Material Figure S2). The treatment, daytime (factor with two levels: day and night) and season (factor with two levels: spring and autumn) factors were used in the third model (spatio-temporal scale). Then, using the *model.sel* function from the 'MuMIn' package [35], this set of models was tested to select the best model(s) based on information criterion corrected for small sample sizes (Akaike Information Criterion—AICc, [36]). The best model was selected as the most parsimonious explanation of the data when Δ AICc was higher than two (Δ AICc > 2) for other models.

3. Results

In total, we recorded 108 attack events based on 1800 observation events (i.e., the predation rate was 6%). Turn was the most common event in 87 cases, followed by scratches or broken parts in 13 cases and 8 relocations. The majority of turns were between 5 and 10° from the original position, and the maximal turn reached 45°. The mean distance for relocation was 11.8 cm and in one case the decoy was taken. Focusing on possible predators, we observed bird droppings (N = 2) and wild boar hair (N = 1) on turned or scratched decoys. Some of the recorded predation events are shown on Figure 1c–g.

The model selection showed that the "spatio-temporal scale" model was the best one, suggesting spatial and temporal constraints are the most influential explanatory variables on predation pressure (Table 2). Therefore, we revealed that the predation pressure was significantly higher in both treatments than in the control forest (Table 3, Figure 2a). Moreover, the pressure was higher during nights than daytime (Figure 2b). On the contrary, no effect of season or environmental variables was confirmed.

Table 2. Summary of the model selection using estimations based on the calculated AICc value of the models, serving as the weight of evidence in favor of the different models. The most parsimonious model (delta < 2) is emphasized in bold.

Model	df	LogLik	AICs	Delta	Weight
Spatio-temporal scale	**5**	**−179.890**	**370.1**	**0.00**	**0.833**
Spatial scale	3	−183.925	374.0	3.86	0.121
Micro-spatial scale	6	−181.705	375.9	5.77	0.046

Table 3. The effects of treatment, daytime, season and environmental variables on the predation pressure. Significant effects are in bold, marginal in italics.

Model	Explanatory Variables	χ2	df	*p*
Spatio-temporal scale	**treatment**	**11.334**	**2**	**0.003**
	daytime	4.444	1	0.035
	season	*3.625*	*1*	*0.056*
Spatial scale	**treatment**	**11.286**	**2**	**0.004**
Micro-spatial scale	treatment	0.203	2	0.903
	leaf litter	0.973	1	0.323
	bare soil	2.404	1	0.121
	tree	0.010	1	0.918

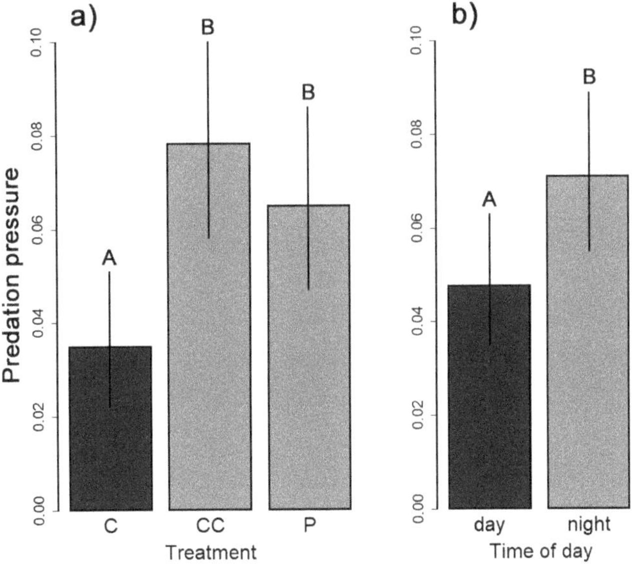

Figure 2. The effect of treatment (**a**) and time of day (**b**) on the predation pressure. Vertical lines represent a 95% confidence interval and different capital letters above the bars indicate significant differences based on Tukey multiple comparisons of means.

4. Discussion

We revealed that 3D-printed decoys represent a suitable approach for testing predation pressure on large carabids, since predators interacted with them. The predation pressure in our study area was affected by spatial as well as temporal constraints. We recorded more attack events in both treatments than in the control. Moreover, the predation pressure on the exposed carabid decoys was higher at night than during the daytime, and it did not correlate with any microhabitat features, such as leaf litter or bush cover, as they are not important for predators of large carabids. Taking into account our previous findings on habitat utilization by large carabids [28], we suggest that both forestry treatments can act as an ecological trap for carabids, as these plots are used only temporarily and with a higher risk of mortality.

Although marks on decoys made from polylactic acid filament cannot deliver a detailed determination of predator identity in comparison with plasticine models [37], we were able to determine some of the predators and suggest other possible candidates. Considering the types of attack, the turn was the most common event, occurring in most cases. Decoy turn may suggest an interest of the potential predator in the exposed decoy, investigating its edibility. Contrarily, scratches and relocations are clear signs of feeding attempts. Wild boar (*Sus scrofa* L., 1758) was determined directly since we found its hair on one relocated decoy. Indeed, various carabid species were recorded as a part of the wild boar's diet [38]. From other mammalian predators, we do not have direct evidence for attacking decoys, but their interest can be presumed as they commonly occur in oak–hornbeam forests of the Pilis Mountains. Eurasian badger (*Meles meles* L., 1758) is already known for feeding on carabids, including large ones, such as *C. coriaceus* [39,40]. Other potential predators could be the northern white-breasted hedgehog (*Erinaceus roumanicus* Martin, 1838) or the greater mouse-eared bat (*Myotis myotis* Borkhausen, 1797), which collect ground-dwelling insects directly from the ground, and large carabids were previously found in its droppings [13]. Moreover, we also recorded bird droppings on two turned decoys. Since we found these droppings exclusively in the spring season, carabids may provide food not only for adult birds but also for nestlings. In addition, Eurasian nuthatch (*Sitta europaea* L., 1758) can feed its young by a high proportion of large carabid species, including the genera *Carabus*, *Calosoma* or *Pterostichus* [12].

The predation pressure was higher in treatments than in the control undisturbed forest, thus the risk of beetles' mortality evidently increased in these treatments. Some large carabids, including *C. coriaceus*, can penetrate clear cuts and preparation cuts in relatively high numbers, likely for foraging, since there is a high competition for the limited resources in the structurally homogeneous undisturbed forest [28]. This kind of habitat utilization is rather transient, since individuals are able to leave the treatment sites within a couple of days [8,28]. There is a high risk of predation in both treatments modified by forestry practices; therefore, these habitats may act as an ecological trap for carabids. The ecological trap is a habitat that is actively selected by individuals despite higher costs of utilization, such as increased mortality and low breeding success [19]. Modified habitat patches are often mentioned as an example of the ecological trap for various animal groups, including insects [19,41]. Our results do not support the increasing disturbance hypothesis [42], suggesting that predation pressure in modified habitats is lower than in undisturbed ones due to the declining number of predators [43,44]. Treatments included in our study were, however, relatively small (with diameter of 80 m) in comparison with landscape gradients and surrounded by undisturbed semi-natural forest. We can presume that large predators, such as wild boars, could easily move across plots regardless of disturbances.

Forestry treatments differ in terms of microclimatic conditions and the structure of understory vegetation from the control forest [29,30]. Nevertheless, we found no effect of any environmental variable on the predation pressure. Microhabitat features, such as the cover of leaf litter or herbs, of a particular patch are important factors for the distribution of carabids [45–47], but not for their predators, as they move at larger scales and likely consider the same patch as structurally homogeneous. Additionally, the predation pressure was higher at night than in the daytime. This seems to be coherent with the fact that *C. coriaceus* is predominantly nocturnal [6,8] and most mammalian predators are more active during nights, which may result in an overlap in activity of prey and potential predators. One of the options to avoid predators is to switch movement activity to the daytime, when the predation pressure is lower. Indeed, beetles, previously considered as strictly nocturnal, were observed to be active during the daytime as well (see [6,8] for activity patterns of *C. coriaceus*). However, there is still a risk of diurnal predators, such as birds. Another option is to minimize the encounters with predators by being hidden as much as possible in shelters. Based on radio-tracking studies, large carabids are able to be inactive for a couple of days, hidden under leaf litter or burrowed in soil [5–7]. In *C. coriaceus*, this anti-predation behavior was observed especially in undisturbed forest [8,28], where a high amount of leaf litter seems to be ideal to hide.

It is also worth mentioning some methodological perspectives and limitations of using 3D-printed decoys for testing predation pressure. This approach is suitable for recording the exploratory behavior of various predators, as they frequently interacted with decoys and turned them. Exact predator identification was, however, limited, as the used printing material (polylactic acid filament) is rather tough for recording soft and small bites. Originally, we experimented with a softer and more elastic material; nevertheless, we could not print narrower body parts, such as legs or antennas. The surface of 3D-printed decoys also could not mimic the structural colors of the real carabid cuticle, especially if the model species has metallic iridescent colorization [14], and this may possibly bias attractiveness for predators. Although we installed decoys far away from each other, there is a chance that the same animals might attack them, leading to an overestimation of predation pressure.

5. Conclusions

Nowadays, ecological studies attempt to focus on the functional aspect of the studied ecosystems [17,48]. Although community measures are widely used, new facets appear and serve as a good proxy for ecosystem functioning including animal behavior. The estimation of predation risk can reflect how animals can select their habitat for foraging, breeding or overwintering. However, for making such a conclusion, the knowledge of the spatio-temporal habitat use of beetles is required, suggesting that other methods should be employed alongside decoys, such as pitfall trapping and radio telemetry. Understanding how interactions between prey and predators change in modified habitats is crucial for better predictions of species-specific responses to habitat alteration. We revealed that habitats modified by forestry practices may act as an ecological trap for carabids. We also proved that the estimation of predation on ground-dwelling predators can be a good proxy for identifying key habitats for conservation [49]. These clues mentioned above may also help us to scrutinize the potential knowledge gaps in understanding how animal behavior can be generalized as a functional component of community-level measures.

Supplementary Materials: The following are available online at https://www.mdpi.com/article/10.3390/d13100484/s1, Figure S1: Three-dimensional model of the *Carabus coriaceus* decoy used for 3D printing. The original file in .stl format (205 MB) can be provided upon request, Figure S2: The overview of the Spearman's rank correlation test for exploring the relationship between microhabitat features: cover of bare soil, leaf litter, herbal layer, shrubbery and the presence of tree.

Author Contributions: Conceptualization, J.R. and Z.E.; methodology, J.R. and Z.E.; formal analysis, J.R. and Z.E.; investigation, J.R. and Z.E.; writing—original draft preparation, J.R. and Z.E.; writing—review and editing, J.R. and Z.E.; visualization, J.R. and Z.E.; funding acquisition, Z.E. All authors have read and agreed to the published version of the manuscript.

Funding: This research was funded by Hungarian Research Fund (NKFIH-K 128441) and by the Hungarian Academy of Sciences (MTA KEP, Ecology for Society project).

Institutional Review Board Statement: Not applicable.

Informed Consent Statement: Not applicable.

Data Availability Statement: The data presented in this study are available on reasonable request from the corresponding author.

Acknowledgments: The authors thank Pilisi Parkerdő Ltd. (Péter Csépányi, Viktor Farkas, Gábor Szenthe, László Simon) for the maintenance of the experimental site. We are also grateful to Jan Chloupek and Alena Vláčilová for modeling and printing 3D decoys as well as Péter Ódor for comments on the manuscript.

Conflicts of Interest: The authors declare no conflict of interest.

References

1. Lövei, G.L.; Sunderland, K.D. Ecology and behavior of ground beetles (Coleoptera: Carabidae). *Annu. Rev. Entomol.* **1996**, *41*, 231–256. [CrossRef]
2. Brandmayr, P.; Bonacci, T.; Giglio, A.; Talarico, F.F.; Brandmayr, T.Z. The evolution of defence mechanisms in carabid beetles: A review. In *Life and Time: The Evolution of Life and Its History*; Casellato, S., Burighel, P., Minelli, A., Eds.; Cleup: Padova, Italy, 2009; pp. 25–43.
3. Sugiura, S. Predators as drivers of insect defenses. *Entomol. Sci.* **2020**, *23*, 316–337. [CrossRef]
4. Giglio, A.; Vommaro, M.L.; Brandmayr, P.; Talarico, F. Pygidial Glands in Carabidae, an Overview of Morphology and Chemical Secretion. *Life* **2021**, *11*, 562. [CrossRef]
5. Niehues, F.J.; Hockmann, P.; Weber, F. Genetics and dynamics of a Carabus auronitens metapopulation in the Westphalian Lowlands (Coleoptera, Carabidae). *Ann. Zool. Fenn.* **1996**, *33*, 85–96.
6. Riecken, U.; Raths, U. Use of radio telemetry for studying dispersal and habitat use of Carabus coriaceus L. *Ann. Zool. Fenn.* **1996**, *33*, 109–116.
7. Růžičková, J.; Veselý, M. Movement activity and habitat use of Carabus ullrichii (Coleoptera: Carabidae): The forest edge as a mating site? *Entomol. Sci.* **2018**, *21*, 76–83. [CrossRef]
8. Elek, Z.; Růžičková, J.; Ódor, P. Individual decisions drive the changes in movement patterns of ground beetles between forestry management types. *Biologia* **2021**, 1–10. [CrossRef]
9. Ferrante, M.; Cacciato, A.L.; Lövei, G.L. Quantifying predation pressure along an urbanisation gradient in Denmark using artificial caterpillars. *Eur. J. Entomol.* **2014**, *111*, 649–654. [CrossRef]
10. Lövei, G.L.; Ferrante, M. A review of the sentinel prey method as a way of quantifying invertebrate predation under field conditions. *Insect Sci.* **2017**, *24*, 528–542. [CrossRef]
11. Boetzl, F.A.; Konle, A.; Krauss, J. Aphid cards—Useful model for assessing predation rates or bias prone nonsense? *J. Appl. Entomol.* **2020**, *144*, 74–80. [CrossRef]
12. Thiele, H.U. *Carabid Beetles in Their Environments*; Springer: Berlin/Heidelberg, Germany, 1997; 369p. [CrossRef]
13. Graclik, A.; Wasielewski, O. Diet composition of Myotis myotis (Chiroptera, Vespertilionidae) in western Poland: Results of fecal analyses. *Turk. J. Zool.* **2012**, *36*, 209–213. [CrossRef]
14. Fukuda, S.; Konuma, J. Using three-dimensional printed models to test for aposematism in a carabid beetle. *Biol. J. Linn. Soc. Lond.* **2019**, *128*, 735–741. [CrossRef]
15. Negro, M.; Vacchiano, G.; Berretti, R.; Chamberlain, D.E.; Palestrini, C.; Motta, R.; Rolando, A. Effects of forest management on ground beetle diversity in alpine beech (Fagus sylvatica L.) stands. *For. Ecol. Manag.* **2014**, *328*, 300–309. [CrossRef]
16. Elek, Z.; Kovács, B.; Aszalós, R.; Boros, G.; Samu, F.; Tinya, F.; Ódor, P. Taxon-specific responses to different forestry treatments in a temperate forest. *Sci. Rep.* **2018**, *8*, 16990. [CrossRef]
17. Elek, Z.; Růžičková, J.; Ódor, P. Functional plasticity of carabids can presume better the changes in community composition than taxon-based descriptors. *Ecol. Appl.* **2021**, e02460. [CrossRef]
18. Bengtsson, J.; Nilsson, S.G.; Franc, A.; Menozzi, P. Biodiversity, disturbances, ecosystem function and management of European forests. *For. Ecol. Manag.* **2000**, *132*, 39–50. [CrossRef]
19. Ewers, R.M.; Didham, R.K. Confounding factors in the detection of species responses to habitat fragmentation. *Biol. Rev.* **2006**, *81*, 117–142. [CrossRef] [PubMed]
20. Schneider, G.; Krauss, J.; Steffan-Dewenter, I. Predation rates on semi-natural grasslands depend on adjacent habitat type. *Basic Appl. Ecol.* **2013**, *14*, 614–621. [CrossRef]
21. González-Gómez, P.L.; Estades, C.F.; Simonetti, J.A. Strengthened insectivory in a temperate fragmented forest. *Oecologia* **2006**, *148*, 137–143. [CrossRef] [PubMed]
22. Belovsky, G.E.; Slade, J.B.; Stockhoff, B.A. Susceptibility to predation for different grasshoppers: An experimental study. *Ecology* **1990**, *71*, 624–634. [CrossRef]
23. Pitt, W.C. Effects of multiple vertebrate predators on grasshopper habitat selection: Trade-offs due to predation risk, foraging, and thermoregulation. *Evol. Ecol.* **1999**, *13*, 499–515. [CrossRef]
24. Bartholomew, A.; El Moghrabi, J. Seasonal preference of darkling beetles (Tenebrionidae) for shrub vegetation due to high temperatures, not predation or food availability. *J. Arid Environ.* **2018**, *156*, 34–40. [CrossRef]
25. Pearson, D.L. The function of multiple anti-predator mechanisms in adult tiger beetles (Coleoptera: Cicindelidae). *Ecol. Entomol.* **1985**, *10*, 65–72. [CrossRef]
26. Sam, K.; Remmel, T.; Molleman, F. Material affects attack rates on dummy caterpillars in tropical forest where arthropod predators dominate: An experiment using clay and dough dummies with green colourants on various plant species. *Entomol. Exp. Appl.* **2015**, *157*, 317–324. [CrossRef]
27. Howe, A.G.; Nachman, G.; Lövei, G.L. Predation pressure in Ugandan cotton fields measured by a sentinel prey method. *Entomol. Exp. Appl.* **2015**, *154*, 161–170. [CrossRef]
28. Růžičková, J.; Bérces, S.; Ackov, S.; Elek, Z. Individual movement of large carabids as a link for activity density patterns in various forestry treatments. *Acta Zool. Acad. Sci. Hung.* **2021**, *67*, 77–86. [CrossRef]

29. Tinya, F.; Kovács, B.; Prättälä, A.; Farkas, P.; Aszalós, R.; Ódor, P. Initial understory response to experimental silvicultural treatments in a temperate oak-dominated forest. *Eur. J. For. Res.* **2019**, *138*, 65–77. [CrossRef]
30. Kovács, B.; Tinya, F.; Guba, E.; Németh, C.; Sass, V.; Bidló, A.; Ódor, P. The short-term effects of experimental forestry treatments on site conditions in an oak–hornbeam forest. *Forests* **2018**, *9*, 406. [CrossRef]
31. Kovács, B.; Tinya, F.; Németh, C.; Ódor, P. Unfolding the effects of different forestry treatments on microclimate in oak forests: Results of a 4-yr experiment. *Ecol. Appl.* **2020**, *30*, e02043. [CrossRef]
32. Effect of Forestry Treatments on Forest Site, Regeneration and Biodiversity. An Experimental Study. Available online: https://piliskiserlet.ecolres.hu/en (accessed on 16 September 2021).
33. Blender—A 3D Modelling and Rendering Package. Available online: http://www.blender.org (accessed on 16 September 2021).
34. R Core Team. *R: A Language and Environment for Statistical Computing*; R Foundation for Statistical Computing: Vienna, Austria, 2019.
35. Bartoń, K. MuMIn: Multi-Model Inference. Available online: https://CRAN.R-project.org/package=MuMIn (accessed on 16 September 2021).
36. Burnham, K.P.; Anderson, D.R. *Model Selection and Multimodel Inference: A Practical Information Theoretic Approach*; Springer: New York, NY, USA, 2002; 488p.
37. Low, P.A.; Sam, K.; McArthur, C.; Posa, M.R.C.; Hochuli, D.F. Determining predator identity from attack marks left in model caterpillars: Guidelines for best practice. *Entomol. Exp. Appl.* **2014**, *152*, 120–126. [CrossRef]
38. Schley, L.; Roper, T.J. Diet of wild boar Sus scrofa in Western Europe, with particular reference to consumption of agricultural crops. *Mammal Rev.* **2003**, *33*, 43–56. [CrossRef]
39. Fletcher, J.G. The Diet and Habitat Utilisation of the Badger (Meles meles) in an Area to the South of Durham City. Ph.D. Thesis, Durham University, Durham, UK, 1992.
40. Marassi, M.; Biancardi, C. Diet of the Eurasian badger (Meles meles) in an area of the Italian Prealps. *Hystrix It. J. Mamm.* **2002**, *13*, 19–28. [CrossRef]
41. Ries, L.; Fagan, W.F. Habitat edges as a potential ecological trap for an insect predator. *Ecol. Entomol.* **2003**, *28*, 567–572. [CrossRef]
42. Gray, J.S. Effects of environmental stress on species rich assemblages. *Biol. J. Linn. Soc. Lond.* **1989**, *37*, 19–32. [CrossRef]
43. Eötvös, C.B.; Lövei, G.L.; Magura, T. Predation pressure on sentinel insect prey along a riverside urbanization gradient in Hungary. *Insects* **2020**, *11*, 97. [CrossRef]
44. Eötvös, C.B.; Magura, T.; Lövei, G.L. A meta-analysis indicates reduced predation pressure with increasing urbanization. *Landsc. Urban Plan.* **2018**, *180*, 54–59. [CrossRef]
45. Niemelä, J.; Spence, J.R.; Spence, D.H. Habitat associations and seasonal activity of ground-beetles (Coleoptera, Carabidae) in central Alberta. *Can. Entomol.* **1992**, *124*, 521–540. [CrossRef]
46. Pearce, J.L.; Venier, L.A.; McKee, J.; Pedlar, J.; McKenney, D. Influence of habitat and microhabitat on carabid (Coleoptera: Carabidae) assemblages in four stand types. *Can. Entomol.* **2003**, *135*, 337–357. [CrossRef]
47. Wehnert, A.; Wagner, S. Niche partitioning in carabids: Single-tree admixtures matter. *Insect Conserv. Divers.* **2019**, *12*, 131–146. [CrossRef]
48. Murray, B.D.; Holland, J.D.; Summerville, K.S.; Dunning, J.B.; Saunders, M.R.; Jenkins, M.A. Functional diversity response to hardwood forest management varies across taxa and spatial scales. *Ecol. Appl.* **2017**, *27*, 1064–1081. [CrossRef] [PubMed]
49. Shochat, E.; Warren, P.S.; Faeth, S.H.; McIntyre, N.E.; Hope, D. From patterns to emerging processes in mechanistic urban ecology. *Trends Ecol. Evol.* **2006**, *21*, 186–191. [CrossRef] [PubMed]

MDPI
St. Alban-Anlage 66
4052 Basel
Switzerland
Tel. +41 61 683 77 34
Fax +41 61 302 89 18
www.mdpi.com

Diversity Editorial Office
E-mail: diversity@mdpi.com
www.mdpi.com/journal/diversity